Table of Contents

History: Einstein and DeBroglie 1

Mathematics of Waves 5

Matter Waves and the Schrödinger Equation 19

    Free Particle 21

    Bound States 30

    Scattering States 49

Hilbert Space and Dirac Notation 66

Quantum Mechanics in Three Dimensions 69

    Laplacian for Spherically Symmetric Potentials 76

    Hydrogen and the Coulomb Potential 84

Part I review 92

Spin: Zeeman, precession, SOI, Clebsch-Gordan 96

Approximation Methods I:

    Time-independent perturbation theory 111

    Variational Principle 123

    Semiclassical/WKB approximation 129

Identical particles, exchange and Pauli exclusion, bandstructure 137

Approximation Methods II:

    Time-dependent perturbation theory 149

    Adiabatic approximation 158

3D Central-Potential Scattering 161

Introduction to Quantum Physics

There is no _magic_ in QM, and in any case this first course will not address any "quantum weirdness" emphasized in popularizations of the subject.

We need only accept that experiments show electrons behave as _matter waves_. The content of this course follows directly from this fact.

Even this unusual idea (that a physical object can be both a particle _and_ a wave) is not unique to QM of electrons!

Particle-Wave Duality : before advent of QM of electron

Electricity and Magnetism: Maxwell's equations + Lorentz force eqn.

Light propagation: 1-d wave eqn for $\vec{E}(z,t) = E_x(z,t)\hat{x}$

$$\nabla^2 \vec{E} = \frac{1}{c^2}\frac{\partial^2 \vec{E}}{\partial t^2} \longrightarrow \frac{\partial^2}{\partial z^2}E_x = \frac{1}{c^2}\frac{\partial^2}{\partial t^2}E_x \qquad c = \frac{1}{\sqrt{\epsilon_0 \mu_0}} \quad \text{"speed of light"}$$

Solution:
$$E_x(z,t) = E_0 e^{i(kz - \omega t)} \qquad \text{"plane wave"}$$

$$-k^2 E_x = -\frac{\omega^2}{c^2}E_x \quad \Rightarrow \quad \omega = kc \quad \text{"dispersion relation"}$$

"radial frequency" $\omega = 2\pi f$, "wavenumber" $k = \frac{2\pi}{\lambda}$ $\Rightarrow \lambda f = c$
("frequency")   ("wavelength")

Interference, diffraction phenomena experimentally show that light is a wave as described above!

Energy and momentum transport in E&M

Fields carry Momentum: $\vec{P} = \mu_0 \varepsilon_0 \vec{S} = \varepsilon_0 (\vec{\mathcal{E}} \times \vec{B})$ ← "Poynting's vector"

For plane waves, $\vec{\mathcal{E}} \perp \vec{B}$ and in phase, $|B| = \frac{|\mathcal{E}|}{c}$

$p = \varepsilon_0 \frac{|\mathcal{E}|^2}{c} = \frac{E}{c}$ ← proportional to Intensity

Fields carry Energy: $E = \frac{1}{2}\left(\varepsilon_0 |\mathcal{E}|^2 + \frac{1}{\mu_0}|B|^2\right) = \varepsilon_0 |\mathcal{E}|^2 = pc$

However:

Experiments done prior to QM (< late 1920s) conclusively show that this prediction of classical E+M (Energy + momentum carried by light prop. to intensity) does not provide the relevant scale in light–matter interactions!

Experiment #1: Photoelectric effect

(1887 H. Hertz)
(1902 Lenard)

Experimental results show that classical prediction is wrong → **frequency** more important than intensity!

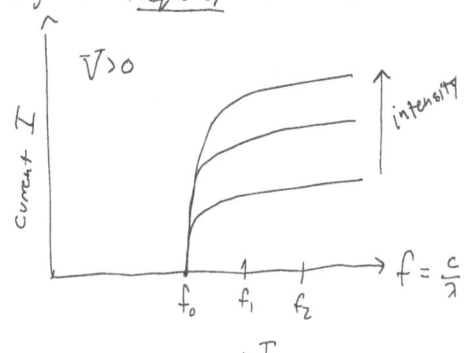

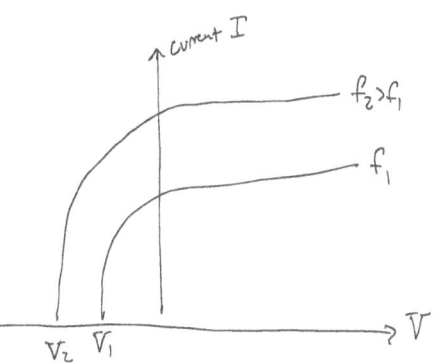

Charge will flow if Energy absorbed from light
$E > W$ ("work function" of metal)
Classical theory predicts we just need to increase the intensity!

Explanation (1905, Einstein)

EM waves come in discrete units of Energy proportional to frequency:

$$E = hf$$ (Note: h has units of Energy·time)

$V < 0$: need $E = hf > W + qV$ for current to flow

$V > 0$: need $E = hf > W$ for current to flow

By repeating the experiment w/ different frequencies f_1 and f_2, measuring thresholds V_1 and V_2, we can solve for h:

$$\begin{aligned} hf_1 &= W + qV_1 \\ hf_2 &= W + qV_2 \end{aligned} \implies h = q\,\frac{V_1 - V_2}{f_1 - f_2}$$

Numerical value same as found for "Blackbody radiation" spectrum (Planck)!

(Note: $E = hf$ is equivalent to $E = \hbar\omega$ where $\hbar = h/2\pi$ and $\omega = 2\pi f$)

Experiment #2: Compton effect (1923)

Classically, (initial) λ = (final) λ'. However, experiment shows $\lambda \neq \lambda'$!

Conservation of Energy and momentum in relativistic regime yields expression

$$\Delta\lambda = \lambda' - \lambda = \frac{h}{mc}(1 - \cos\theta)$$

independent of intensity, matching experimental results!

Conclusion: "light" is made up of particles ("photons") with discrete Energy $E = hf$ and discrete momentum $p = \frac{E}{c} = \frac{hf}{c} = \frac{h}{\lambda}$

So particle-wave duality is not unique to electrons!

Interference

Convincing evidence of wave nature of <u>free</u> electrons is provided by interference / diffraction (more on this later, including in-class demo...)

But what about when electrons are <u>confined</u> e.g. in an atom?

$\Rightarrow$ When waves on a string are confined, <u>discrete modes</u> result.

$$K_n = \frac{n\pi}{L}$$

Can we see the effects of discrete electron wave modes??

Yes! Atomic gases emit photons of well-defined wavelength (and hence energy) when excited by collisions due to high voltage in discharge tubes: $\longrightarrow$

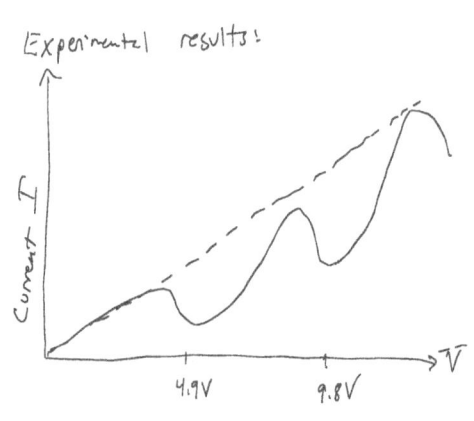

Discharges in the low-pressure gas filled tube are sources of light, which undergo refraction on a prism. We see the line spectrum of the gas.

Further Evidence of discrete electron wave modes : Franck–G. Hertz (1914)

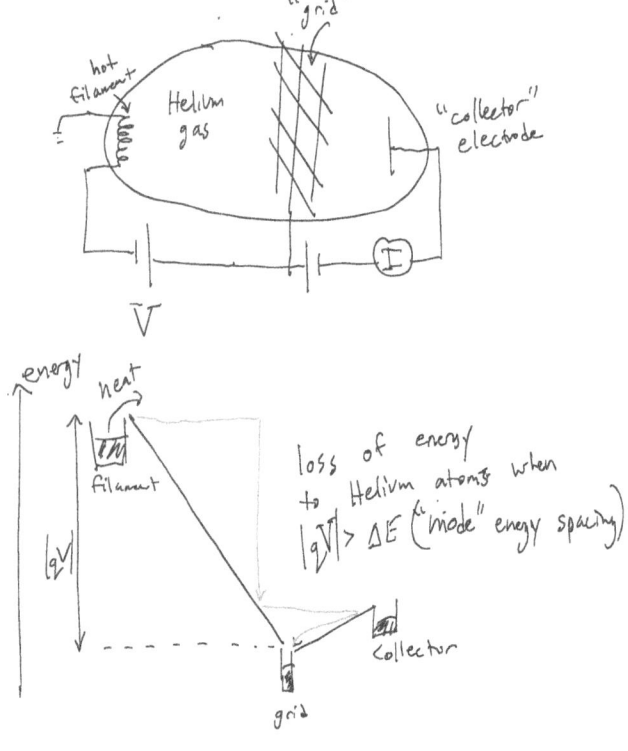

loss of energy to Helium atoms when $|qV| > \Delta E$ ("mode" energy spacing)

So electrons in Helium atom can only absorb discrete units of the free electron's kinetic energy, just like discrete modes of mechanical waves on a string!

If the electron is a wave, then...

1. What is the wave equation it satisfies?

2. What is the corresponding <u>dispersion relation</u>?

3. What is the physical <u>interpretation</u> of this wave?

In this course, we will address these questions, then use their answers to solve basic problems in "quantum mechanics"!

Review: Differential Equations

Ordinary differential equations: Solution is a <u>function</u> of one variable

"special examples": Can be solved by inspection

$$\frac{df(x)}{dx} = C \longrightarrow f(x) = Cx + A$$

$$\frac{df(x)}{dx} = Cf(x) \longrightarrow f(x) = Ae^{Cx}$$

"first order"
<u>one</u> undetermined constant in general sol'n

$$\frac{d^2 f(x)}{dx^2} = C \longrightarrow f(x) = \frac{C}{2} x^2 + A_1 x + A_2$$

$$\frac{d^2 f(x)}{dx^2} = -C^2 f(x) \longrightarrow f(x) = A_1 \cos(Cx) + A_2 \sin(Cx)$$

$$\underset{\substack{\text{(mathematically} \\ \text{equivalent)}}}{\text{OR}} = B_1 e^{iCx} + B_2 e^{-iCx} \qquad (i = \sqrt{-1})$$

"Second order"
two undetermined constants

Complex Numbers

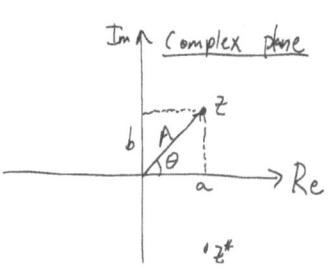

$z = a + ib$ (Cartesian)
$\quad = A e^{i\theta}$ (polar)
$z^* = a - ib$ "Complex conjugate"
$\quad = A e^{-i\theta}$

Connection between polar and Cartesian:
$$e^{i\theta} = \cos\theta + i\sin\theta \qquad \text{"Euler's formula"}$$

proof: Taylor series $\quad f(\theta) = f(0) + f'(0)\theta + \frac{f''(0)}{2!}\theta^2 + \frac{f'''(0)}{3!}\theta^3 + \ldots$

So $\quad f(\theta) = e^{i\theta} = 1 + i\theta - \frac{\theta^2}{2!} + i\frac{\theta^3}{3!} + \ldots = \left(1 - \frac{\theta^2}{2!} + \ldots\right) + i\left(\theta - \frac{\theta^3}{3!} + \ldots\right) = \cos\theta + i\sin\theta$

Consequences:
$$\cos\theta = \frac{e^{i\theta} + e^{-i\theta}}{2} \qquad \sin\theta = \frac{e^{i\theta} - e^{-i\theta}}{2i}$$

$$e^{i\pi} + 1 = 0 \qquad \text{So "beautiful"!}$$

Solution by "brute force"

Example:
$$\frac{df}{dx} = Cf \qquad \text{guess: } f(x) = A_0 + A_1 x + A_2 x^2 + \ldots$$

Substitute:
$$A_1 + 2A_2 x + 3A_3 x^2 + \ldots = CA_0 + CA_1 x + CA_2 x^2 + \ldots$$

Since this must be true for all x, equate like powers of x:

$(x^0): \quad A_1 = CA_0$
$(x^1): \quad 2A_2 = CA_1 = C^2 A_0 \quad \to \quad A_2 = \frac{A_0 C^2}{2}$
$(x^2): \quad 3A_3 = CA_2 = \frac{C^3 A_0}{2} \quad \to \quad A_3 = \frac{A_0 C^3}{3 \cdot 2}$
$\vdots$

So:
$$f(x) = A_0 + A_0 C x + A_0 \frac{C^2 x^2}{2} + A_0 \frac{C^3 x^3}{3 \cdot 2} + \ldots$$
$$= A_0 \left[1 + (Cx) + \frac{(Cx)^2}{2!} + \frac{(Cx)^3}{3!} + \ldots \right] = A_0 e^{Cx}$$

Compare w/ sol'n by inspection!

A simpler route to solution by "brute force"

Solve $f' \equiv \frac{df}{dx} = cf$ by substituting infinite sum $f = \sum_{n=0}^{\infty} A_n x^n$

$$\sum_{n=0}^{\infty} n A_n x^{n-1} = c \sum_{n=0}^{\infty} A_n x^n$$

$$\sum_{n=1}^{\infty} n A_n x^{n-1} = \sum_{n=0}^{\infty} c A_n x^n$$

$$\sum_{n=0}^{\infty} (n+1) A_{n+1} x^n = \sum_{n=0}^{\infty} c A_n x^n$$

$$A_{n+1} = \frac{c}{n+1} A_n \quad \text{"recursion relation"}$$

$$f = A_0 \sum_{n=0}^{\infty} \frac{c^n}{n!} x^n = A_0 \sum_{n=0}^{\infty} \frac{(cx)^n}{n!} = A_0 e^{cx} \quad \checkmark$$

Partial differential equations

Sol'ns are functions of <u>more</u> than one variable

Example: $\frac{\partial}{\partial t} f(x,t) = D \frac{\partial^2}{\partial x^2} f(x,t)$ "Heat"/"Diffusion" equation

[Shorthand notation: $\dot{f} = D f''$]

Always try "separation of variables":

guess: $f(x,t) = X(x) T(t)$

substitute:

$$\frac{X \dot{T}}{D X T} = \frac{D T X''}{D T X}$$

$$\frac{\dot{T}}{D T} = \frac{X''}{X} = -k^2 \quad \text{← constant, units distance}^{-2}$$

↑ only depends on t ↑ only depends on x

2nd order ODE:
$$X'' = -k^2 X$$
$$X(x) = B(k) e^{ikx} \quad (-\infty < k < \infty)$$

1st order ODE:
$$\dot{T} = -D k^2 T$$
$$T(t) = C(k) e^{-Dk^2 t}$$

general sol'n:
$$f(x,t) = \frac{1}{\sqrt{2\pi}} \int_{-\infty}^{\infty} A(k) e^{-Dk^2 t} e^{ikx} dk$$

Initial conditions:
$$f(x, t=0) = \frac{1}{\sqrt{2\pi}} \int_{-\infty}^{\infty} A(k) e^{ikx} dk$$

But how to determine $A(k)$ from $f(x, t=0)$??

Application of initial conditions: "Fourier Transform"

Change of basis in an orthonormal function space

Analogy with Euclidean vector space:

<u>Vectors</u>: $\vec{F} = \sum_i A_i \vec{v}_i$ ($\vec{v}_i$'s are "orthonormal basis vectors")

(2D) example: $\vec{F} = 2\hat{x} + 3\hat{y}$
               A_1 A_2
               $\vec{v}_1$ $\vec{v}_2$

<u>functions</u>: $F(x) = \int_{-\infty}^{\infty} A(k) u_k(x) dk$ ($u_k(x)$'s are "orthonormal basis functions")

$\left(u_k(x) = \dfrac{e^{ikx}}{\sqrt{2\pi}} \text{ for Fourier Transform} \right)$

Determine coefficients

<u>Vectors</u> $\vec{F} = \sum_i A_i \vec{v}_i$

"inner product": $\vec{v}_j \cdot \vec{F} = \vec{v}_j \cdot \sum_i A_i \vec{v}_i = \sum_i A_i (\vec{v}_j \cdot \vec{v}_i)$

$\phantom{\text{"inner product": } \vec{v}_j \cdot \vec{F}} = \sum_i A_i \delta_{ij}$ ← "Kronecker delta" $= 0$ if $i \neq j$
$\phantom{\text{"inner product": } \vec{v}_j \cdot \vec{F} = \sum_i A_i \delta_{ij}}= 1$ $i = j$

$\phantom{\text{"inner product": } \vec{v}_j \cdot \vec{F}} = A_j$

$\Rightarrow A_i = \vec{v}_i \cdot \vec{F}$

Example: $A_1 = \vec{v}_1 \cdot \vec{F} = \hat{x} \cdot (2\hat{x} + 3\hat{y}) = 2(\hat{x} \cdot \hat{x})^1 + 3(\hat{x} \cdot \hat{y})^0 = 2$

We can also find coefficients in a <u>different</u> basis! for example,

if $\vec{v}_1 = \dfrac{\hat{x}+\hat{y}}{\sqrt{2}}$, $\vec{v}_2 = \dfrac{\hat{x}-\hat{y}}{\sqrt{2}}$, Then in this basis,

$A_1 = \vec{v}_1 \cdot \vec{F} = \left(\dfrac{\hat{x}+\hat{y}}{\sqrt{2}}\right)(2\hat{x}+3\hat{y}) = \dfrac{5}{\sqrt{2}}$, $A_2 = \vec{v}_2 \cdot \vec{F} = \dfrac{\hat{x}-\hat{y}}{\sqrt{2}}(2\hat{x}+3\hat{y}) = -\dfrac{1}{\sqrt{2}}$

Determine coefficients

functions $F(x) = \int_{-\infty}^{\infty} A(k) u_k(x) dk$

take inner product w/ another basis function:

$$\int_{-\infty}^{+\infty} u_{k'}^*(x) F(x) dx = \int_{-\infty}^{+\infty} u_{k'}^*(x) \left[\int_{-\infty}^{\infty} A(k) u_k(x) dk \right] dx$$

$$= \int_{-\infty}^{\infty} A(k) \left[\int_{-\infty}^{\infty} u_{k'}^*(x) u_k(x) dx \right] dk$$

$$= \int_{-\infty}^{\infty} A(k) \, \delta(k-k') dk \qquad \left(\begin{array}{l} \text{"Dirac delta fn":} \\ \delta(k-k') = 0 \quad k \neq k' \\ \text{But } \int_{-\infty}^{\infty} \delta(k-k') dk = 1 \end{array} \right)$$

$$= A(k')$$

$$\Rightarrow A(k) = \int_{-\infty}^{\infty} u_k^*(x) F(x) dx \qquad \text{"Fourier transform" of } F(x)$$

This formula now allows us to solve our P.D.E., given initial conditions!

Diffusion equation, revisited

$$\frac{\partial f}{\partial t} = D \frac{\partial^2 f}{\partial x^2} \qquad \text{solution: use F.T.} \qquad f(x,t) = \frac{1}{\sqrt{2\pi}} \int_{-\infty}^{\infty} \tilde{f}(k,t) e^{ikx} dk$$

$$\int_{-\infty}^{\infty} \frac{e^{-ik'x}}{\sqrt{2\pi}} \left[\frac{1}{\sqrt{2\pi}} \int_{-\infty}^{\infty} \dot{\tilde{f}}(k,t) e^{ikx} dk \right] dx = \int_{-\infty}^{\infty} \frac{e^{-ik'x}}{\sqrt{2\pi}} \left[\frac{-D}{\sqrt{2\pi}} \int_{-\infty}^{\infty} k^2 \tilde{f}(k,t) e^{ikx} dk \right] dx$$

$$\int_{-\infty}^{\infty} \dot{\tilde{f}}(k,t) \left[\int_{-\infty}^{\infty} \frac{e^{-ik'x}}{\sqrt{2\pi}} \frac{e^{ikx}}{\sqrt{2\pi}} dx \right] dk = -D \int_{-\infty}^{\infty} k^2 \tilde{f}(k,t) \left[\int_{-\infty}^{\infty} \frac{e^{-ik'x}}{\sqrt{2\pi}} \frac{e^{ikx}}{\sqrt{2\pi}} dx \right] dk$$

$$\int_{-\infty}^{\infty} \dot{\tilde{f}}(k,t) \delta(k-k') dk = -D \int_{-\infty}^{\infty} k^2 \tilde{f}(k,t) \delta(k-k') dk$$

$$\dot{\tilde{f}}(k',t) = -D k'^2 \tilde{f}(k',t) \qquad \text{an O.D.E we know how to solve!}$$

$$\tilde{f}(k,t) = A(k) e^{-Dk^2 t}$$

Then,
$$f(x,t) = \frac{1}{\sqrt{2\pi}} \int_{-\infty}^{\infty} A(k) e^{-Dk^2 t} e^{ikx} dk \qquad \text{Same as solution by separation of variables!}$$

Closed-form solution to diffusion equation

$$f(x, t=0) = \frac{1}{\sqrt{2\pi}} \int_{-\infty}^{\infty} A(k) e^{ikx} dk \xleftrightarrow{\text{Fourier Transform}} A(k) = \frac{1}{\sqrt{2\pi}} \int_{-\infty}^{\infty} f(x, t=0) e^{-ikx} dx$$

Example: $f(x, t=0) = \delta(x)$ "impulse" initial conditions

$$A(k) = \frac{1}{\sqrt{2\pi}} \int_{-\infty}^{\infty} \delta(x) e^{-ikx} dx = \frac{1}{\sqrt{2\pi}}$$

So,

$$f(x,t) = \frac{1}{\sqrt{2\pi}} \int_{-\infty}^{\infty} A(k) e^{-Dk^2 t} e^{ikx} dk = \frac{1}{2\pi} \int_{-\infty}^{\infty} e^{-Dk^2 t} e^{ikx} dk = \frac{1}{2\pi} \int_{-\infty}^{\infty} e^{-Dt\left(k^2 - \frac{ikx}{Dt}\right)} dk$$

"Complete the square":
$$= \frac{1}{2\pi} \int_{-\infty}^{\infty} e^{-Dt\left(k - \frac{ix}{2Dt}\right)^2} e^{-\frac{x^2}{4Dt}} dk$$

$$= \frac{1}{2\pi} e^{-\frac{x^2}{4Dt}} \int_{-\infty}^{\infty} e^{-Dt\left(k - \frac{ix}{2Dt}\right)^2} dk$$

transform variables $y = \sqrt{Dt}\left(k - \frac{ix}{2Dt}\right)$, $dk = \frac{dy}{\sqrt{Dt}}$:
$$= \frac{1}{2\pi} e^{-\frac{x^2}{4Dt}} \left[\underbrace{\int_{-\infty}^{\infty} e^{-y^2} dy}_{\sqrt{\pi} \text{ (see next page)}}\right] \frac{1}{\sqrt{Dt}}$$

Integrating the gaussian

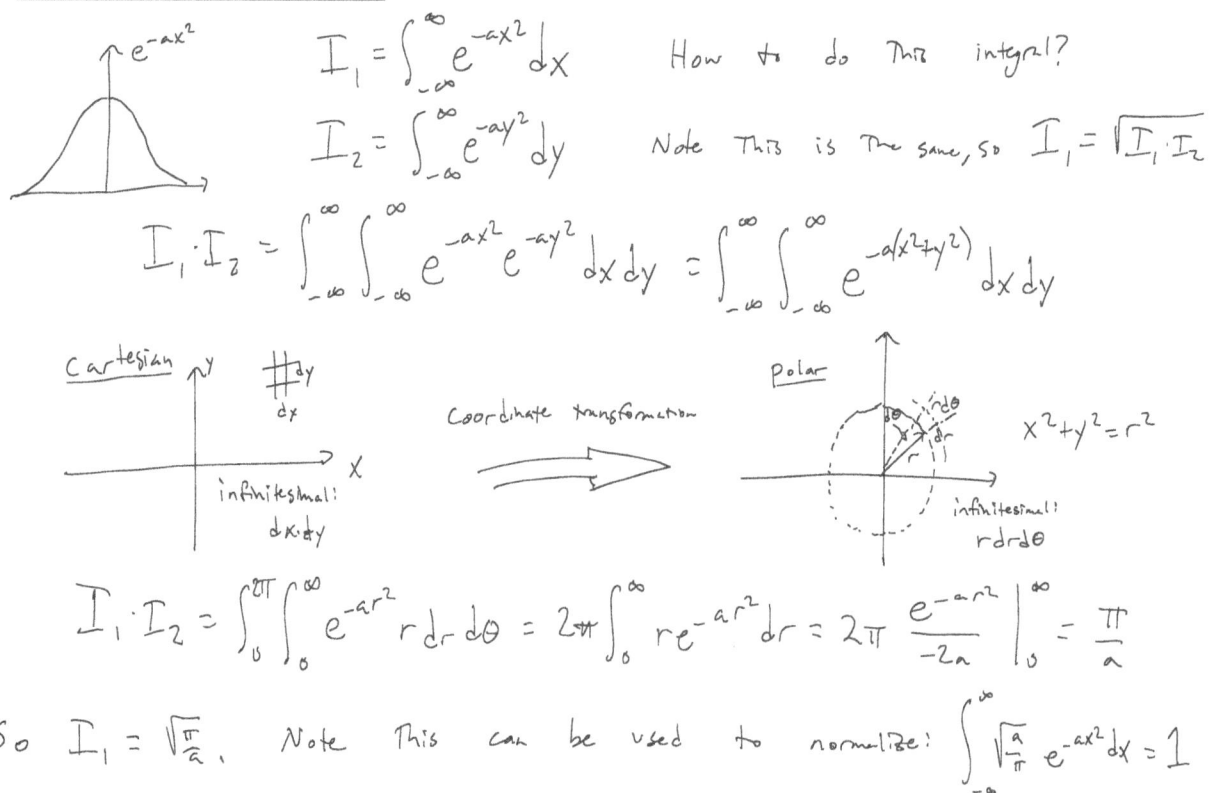

$I_1 = \int_{-\infty}^{\infty} e^{-ax^2} dx$ How to do this integral?

$I_2 = \int_{-\infty}^{\infty} e^{-ay^2} dy$ Note this is the same, so $I_1 = \sqrt{I_1 \cdot I_2}$

$$I_1 \cdot I_2 = \int_{-\infty}^{\infty} \int_{-\infty}^{\infty} e^{-ax^2} e^{-ay^2} dx\, dy = \int_{-\infty}^{\infty} \int_{-\infty}^{\infty} e^{-a(x^2+y^2)} dx\, dy$$

Cartesian → Coordinate transformation → Polar $x^2 + y^2 = r^2$

infinitesimal: $dx \cdot dy$ infinitesimal: $r\, dr\, d\theta$

$$I_1 \cdot I_2 = \int_0^{2\pi} \int_0^{\infty} e^{-ar^2} r\, dr\, d\theta = 2\pi \int_0^{\infty} r e^{-ar^2} dr = 2\pi \left. \frac{e^{-ar^2}}{-2a} \right|_0^{\infty} = \frac{\pi}{a}$$

So $I_1 = \sqrt{\frac{\pi}{a}}$. Note this can be used to normalize: $\int_{-\infty}^{\infty} \sqrt{\frac{a}{\pi}} e^{-ax^2} dx = 1$

Solution to diffusion equation with impulse initial conditions

$$f(x,t) = \frac{1}{2\sqrt{\pi D t}} e^{-\frac{x^2}{4Dt}}$$ Normalized "Gaussian" in x

At $t=0$ ($\delta(x)$, infinitesimally narrow but $\int_{-\infty}^{\infty} f(x)\,dx = 1$)

As t grows, distribution gets wider in space Δx and amplitude decreases to maintain area under curve constant.

Standard deviation $\Delta x = \sqrt{2Dt}$

$t_2 > t_1$

In general, normalized gaussian has form $f(x) = \frac{1}{\Delta x \sqrt{2\pi}} e^{-\frac{x^2}{2\Delta x^2}}$ such that

$$\Delta x = \sqrt{\overline{\Delta x^2}} = \sqrt{\int_{-\infty}^{\infty} (x - \langle x \rangle)^2 f(x)\, dx}$$ (note that $\langle x \rangle = 0$)

Fourier transform of gaussian

$\left(f(x) = e^{-ax^2},\ \Delta x = \frac{1}{\sqrt{2a}} \right)$

$$A(k) = \int_{-\infty}^{\infty} e^{-ax^2} \frac{e^{-ikx}}{\sqrt{2\pi}}\, dx = \frac{1}{\sqrt{2\pi}} \int_{-\infty}^{\infty} e^{-a\left(x^2 + \frac{ikx}{a}\right)} dx$$

Complete the square:
$$= \frac{1}{\sqrt{2\pi}} \int_{-\infty}^{\infty} e^{-a\left(x + \frac{ik}{2a}\right)^2} e^{-\frac{k^2}{4a}}\, dx$$

Variable transformation: $z = \sqrt{a}\left(x + \frac{ik}{2a}\right),\ dx = \frac{dz}{\sqrt{a}}$

$$= \frac{1}{\sqrt{2\pi a}} e^{-\frac{k^2}{4a}} \overbrace{\int_{-\infty}^{\infty} e^{-z^2}\, dz}^{\sqrt{\pi}} = \frac{1}{\sqrt{2a}} e^{-\frac{k^2}{4a}}$$

So, the F.T. of gaussian in x is a gaussian in k!

"Heisenberg uncertainty principle"

Note the reciprocal relationship between the standard deviation "uncertainty" $\Delta x = \frac{1}{\sqrt{2a}}$ of $f(x)$ and its transform $\Delta k = \sqrt{2a}$ of $A(k)$. A large Δx means a small Δk and vice versa. Therefore,

$$\Delta x \cdot \Delta k = 1$$

The gaussian minimizes this product, so in general $\Delta x \Delta k \geq 1$. Physicists call this the "Heisenberg uncertainty principle" in the context of QM, but clearly it is only a consequence of the mathematics of the Fourier Transform (and its wavelike basis fns)!

Generality of uncertainty principle

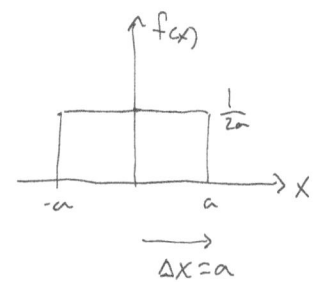

$$A(k) = \int_{-\infty}^{\infty} f(x) \frac{e^{-ikx}}{\sqrt{2\pi}} dx = \int_{-a}^{a} \frac{1}{2a} \frac{e^{-ikx}}{\sqrt{2\pi}} dx = \frac{1}{2a\sqrt{2\pi}} \frac{e^{-ikx}}{-ik} \Big|_{-a}^{a}$$

$$= \frac{1}{2a\sqrt{2\pi}} \frac{e^{-ika} - e^{ika}}{-ik} = \frac{1}{ka\sqrt{2\pi}} \frac{(e^{ika} - e^{-ika})}{2i} = \frac{1}{\sqrt{2\pi}} \frac{\sin ka}{ka}$$

$$\propto \text{sinc}(ka)$$

$\Delta x = a$, $\Delta k = \frac{\pi}{a}$! $\Delta x \Delta k = \pi > 1$

So reciprocal relationship holds!

Experimenting with Fourier transforms

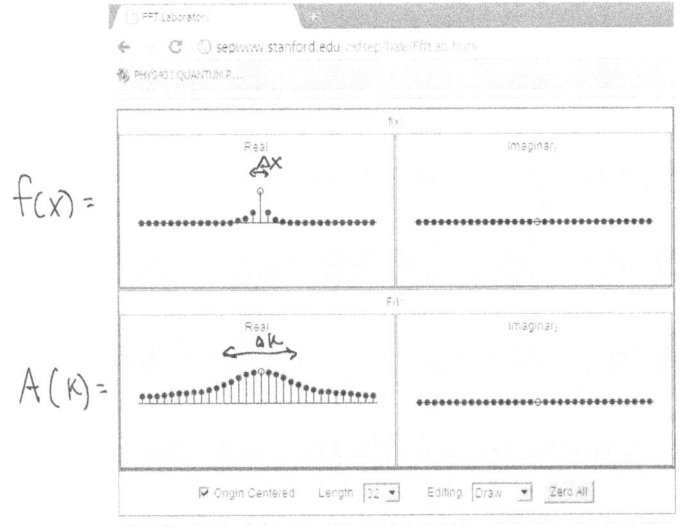

$f(x) =$

$A(k) =$

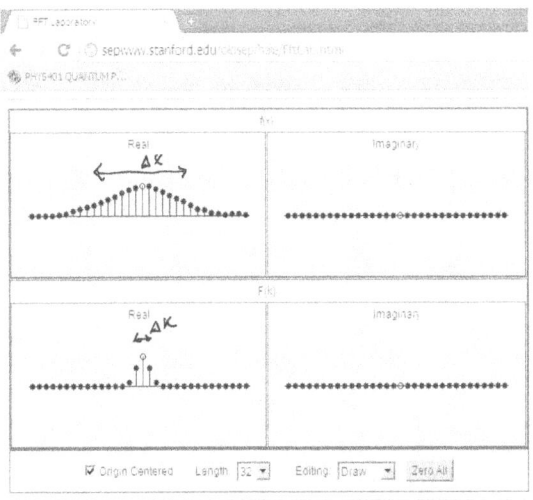

Note reciprocal relationship between Δx and Δk!

Boundary conditions: eigenvalue/eigenfunction problems

example: Classical wave equation

$$\frac{\partial^2 f}{\partial x^2} = \frac{1}{c^2}\frac{\partial^2 f}{\partial t^2}$$

String, $x=0$, $f(x=0, t) = 0$; $x=L$, $f(x=L, t) = 0$

Separation of variables: $f(x,t) = X(x) T(t)$

$$\frac{X''T}{XT} = \frac{1}{c^2}\frac{X\ddot{T}}{XT}$$

$$\frac{X''}{X} = \frac{\ddot{T}}{c^2 T} = -k^2$$

$X'' = -k^2 X$ and $\ddot{T} = -k^2 c^2 T$

$\downarrow$

$X(x) = A\cos kx + B\sin kx$

Apply Boundary Conditions:

$X(x=0) = 0 = A$

$X(x=L) = B\sin kL = 0$

So $kL = n\pi$, $n = 1, 2, 3, \ldots$

$k = \frac{n\pi}{L}$ (discrete) "eigenvalues"

Then, our general solution can be written

$$f(x,t) = \sqrt{\frac{2}{L}} \sum_n \sin k_n x \cdot \left[C_n \sin \omega_n t + D_n \cos \omega_n t \right]$$

where $\omega_n = k_n c$

Applying initial conditions

With $f(x, t=0)$ specified, we can determine the coefficients C_n, D_n by using orthonormality of the basis functions ("modes"),

e.g. $$f(x, t=0) = \sqrt{\frac{2}{L}} \sum_n D_n \sin k_n x$$

Take inner product w/ orthonormal basis functions $u_m(x) = \sqrt{\frac{2}{L}} \sin k_m x$:

$$\sqrt{\frac{2}{L}} \int_0^L \sin k_m x \, f(x, t=0) \, dx = \int_0^L \frac{2}{L} \sin k_m x \left(\sum_n D_n \sin k_n x \right) dx$$

Reverse order of sum/integral $= \sum_n D_n \left(\int_0^L \frac{2}{L} \sin k_m x \sin k_n x \right) dx$

By orthonormality, $= \sum_n D_n \, \delta_{nm} = D_m$

So, $D_n = \int_0^L u_n^*(x) \, f(x, t=0) \, dx$ look familiar?

Numerical solution of $\frac{d^2 X(x)}{dx^2} = -k^2 X(x)$

Use symmetric definition of derivative $\frac{d}{dx} X(x) = \lim_{\Delta x \to 0} \frac{X(x + \frac{\Delta x}{2}) - X(x - \frac{\Delta x}{2})}{\Delta x}$

Then, $\frac{d^2}{dx^2} X(x) = \lim_{\Delta x \to 0} \frac{\frac{X(x+\Delta x) - X(x)}{\Delta x} - \frac{X(x) - X(x-\Delta x)}{\Delta x}}{\Delta x} = \lim_{\Delta x \to 0} \frac{X(x - \Delta x) - 2 X(x) + X(x + \Delta x)}{\Delta x^2}$

Now, instead of passing to limit $\Delta x \to 0$ to recover continuous differential operator, use "Finite differences": Approximate derivatives by using nonzero value of Δx i.e. discretize continuous variable x: N segments from $x=0$ to $x=L$:

Then,
$$\frac{d^2 X(x_i)}{dx^2} \sim \frac{X(x_{i-1}) - 2 X(x_i) + X(x_{i+1})}{\Delta x^2} \equiv \boxed{\frac{X_{i-1} - 2 X_i + X_{i+1}}{\Delta x^2} = -k^2 X_i}$$

From differential equation to system of linear algebraic equations

Evaluate at every value of x_i using finite differences approximation:

$i=1:\ \frac{1}{\Delta x^2}\left(-2X_1 + X_2\right) = -k^2 X_1$

$i=2:\ \frac{1}{\Delta x^2}\left(X_1 - 2X_2 + X_3\right) = -k^2 X_2$

$i=3:\ \frac{1}{\Delta x^2}\left(X_2 - 2X_3 + X_4\right) = -k^2 X_3$

$\vdots$

$i=N:\ \frac{1}{\Delta x^2}\left(X_{N-1} - 2X_N\right) = -k^2 X_N$

Equivalent to:

$$\frac{1}{\Delta x^2}\begin{bmatrix} -2 & 1 & 0 & 0 & \cdots \\ 1 & -2 & 1 & 0 & \cdots \\ 0 & 1 & -2 & 1 & 0 & \cdots \\ & & & \ddots & \\ & & \cdots & 0 & 1 & -2 \end{bmatrix} \begin{bmatrix} X_1 \\ X_2 \\ X_3 \\ \vdots \\ X_N \end{bmatrix} = -k^2 \begin{bmatrix} X_1 \\ X_2 \\ X_3 \\ \vdots \\ X_N \end{bmatrix}$$

a **matrix** eigenvalue problem!

Calculating eigenvalues and eigenvectors

$\overleftrightarrow{M}\vec{x} = \lambda \vec{x} = \lambda \overleftrightarrow{I}\vec{x}$ identity $\overleftrightarrow{I} = \begin{bmatrix} 1 & 0 & \\ 0 & 1 & 0 \\ 0 & 0 & \ddots \end{bmatrix}$ (diagonal)

$\left(\overleftrightarrow{M} - \lambda \overleftrightarrow{I}\right)\vec{x} = 0$

$\det\left(\overleftrightarrow{M} - \lambda \overleftrightarrow{I}\right) = 0$

Example: $\overleftrightarrow{M} = \begin{bmatrix} 0 & 1 \\ 1 & 0 \end{bmatrix}$ $\Rightarrow$ $\det\left(\begin{bmatrix} 0 & 1 \\ 1 & 0 \end{bmatrix} - \begin{bmatrix} \lambda & 0 \\ 0 & \lambda \end{bmatrix}\right) = \det\begin{bmatrix} -\lambda & 1 \\ 1 & -\lambda \end{bmatrix} = \lambda^2 - 1 = 0$

$\lambda = +1, -1$

eigenvectors:

$\lambda = +1:\ \begin{bmatrix} -1 & 1 \\ 1 & -1 \end{bmatrix}\begin{bmatrix} x_1 \\ x_2 \end{bmatrix} = \begin{bmatrix} 0 \\ 0 \end{bmatrix}$ $x_1 = x_2$ $\vec{x} = \begin{bmatrix} 1/\sqrt{2} \\ 1/\sqrt{2} \end{bmatrix}$ (elements chosen to normalize $|\vec{x}| = 1$)

$\lambda = -1:\ \begin{bmatrix} 1 & 1 \\ 1 & 1 \end{bmatrix}\begin{bmatrix} x_1 \\ x_2 \end{bmatrix} = \begin{bmatrix} 0 \\ 0 \end{bmatrix}$ $x_1 = -x_2$ $\vec{x} = \begin{bmatrix} 1/\sqrt{2} \\ -1/\sqrt{2} \end{bmatrix}$

But if eigenvalues are roots of a polynomial, in general they are complex valued. Since eigenvalues have physical meaning in our problem, how can we be certain they are always real valued?

Matrix Symmetry

The key to physically meaningful eigenvalues/eigenvectors is that we will always have a "Hermitian" matrix: One that is equal to its own Hermitian conjugate (transpose/complex conjugate): $\overleftrightarrow{H} = \overleftrightarrow{H}^\dagger$

(Note: $(AB)^\dagger = B^\dagger A^\dagger$)

Proof that eigenvalues of Hermitian matrix are always real:

$$\overleftrightarrow{H}\vec{x} = \lambda \vec{x}$$

$$(\overleftrightarrow{H}\vec{x})^\dagger = (\lambda \vec{x})^\dagger$$

$$\vec{x}^\dagger \overleftrightarrow{H}^\dagger = \vec{x}^\dagger \lambda^*$$

$$\vec{x}^\dagger \overleftrightarrow{H} = \vec{x}^\dagger \lambda^*$$

$$\vec{x}^\dagger \overleftrightarrow{H} \vec{x} = \vec{x}^\dagger \lambda^* \vec{x}$$

$$\overleftrightarrow{H}\vec{x} = \lambda^* \vec{x} \quad \longrightarrow \quad \text{So } \lambda = \lambda^* \text{ and therefore must be real!}$$

Eigenvectors of Hermitian matrices

Take two different eigenvector/eigenvalue pairs of hermitian matrix H:

$$\overleftrightarrow{H}\vec{x}_1 = a\vec{x}_1 \qquad \overleftrightarrow{H}\vec{x}_2 = b\vec{x}_2$$

$$\vec{x}_2^\dagger \overleftrightarrow{H} \vec{x}_1 = \vec{x}_2^\dagger a \vec{x}_1 \qquad (\overleftrightarrow{H}\vec{x}_2)^\dagger = (b\vec{x}_2)^\dagger$$

$$\vec{x}_2^\dagger \overleftrightarrow{H} \vec{x}_1 = a\vec{x}_2^\dagger \vec{x}_1 \qquad \vec{x}_2^\dagger \overleftrightarrow{H}^\dagger = \vec{x}_2^\dagger b^*$$

$$\vec{x}_2^\dagger \overleftrightarrow{H}^\dagger \vec{x}_1 = \vec{x}_2^\dagger b^* \vec{x}_1 = b^* \vec{x}_2^\dagger \vec{x}_1$$

So: $a\vec{x}_2^\dagger \vec{x}_1 = b^* \vec{x}_2^\dagger \vec{x}_1$ Since a, b are real and distinct, this can only be true if $\vec{x}_2^\dagger \vec{x}_1 = 0$ $(\vec{x}_2 \cdot \vec{x}_1 = 0)$

In other words, the eigenvectors of a Hermitian matrix are orthogonal!

Question: What are the eigenvectors of this matrix?

$$\begin{bmatrix} -2 & 1 & 0 \\ 1 & -2 & 1 \\ 0 & 1 & -2 \end{bmatrix}$$

We have seen this matrix before: It was the matrix equivalent of the second derivative from the wave equation! So its eigenvectors are approximations to the modes of a standing wave satisfying $\overleftrightarrow{H}\vec{x} = \lambda \vec{x}$:

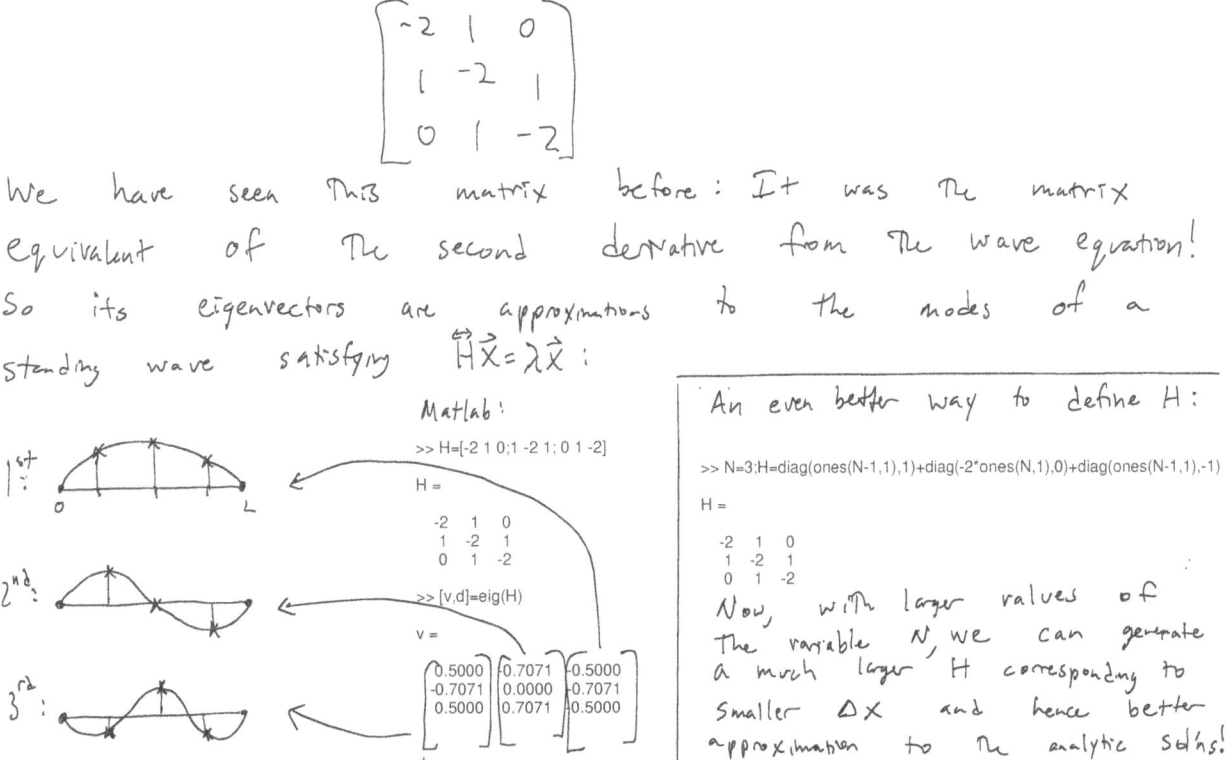

Numerical solution to standing wave eigenvalue problem

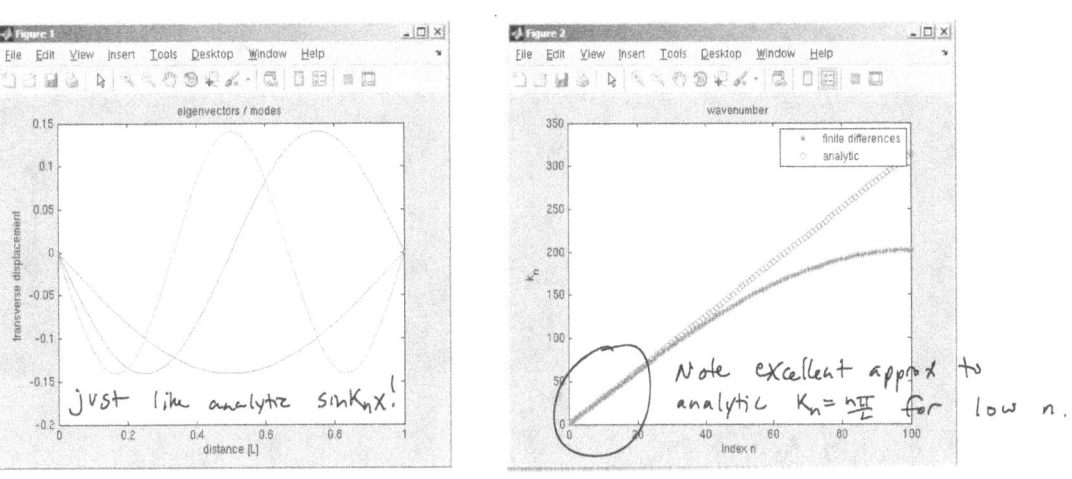

Matlab hints!

$[v, d] = \text{eig}(H)$ returns eigenvectors as columns in the matrix v: The nth eigenvector is $v(:,n)$. The corresponding eigenvalues are given as diagonal elements of matrix d. They can be recovered into an array via $\text{diag}(d)$.

Using MATLAB functions

1. Create a text file with extension ".m" in your path
2. In the first line, define
    function <output> = <function name>(<input>)
3. In subsequent lines, calculate output.

Example: a text file called 'genH.m':

```
function H=genH(N)
H=diag(-2*ones(N,1))+diag(ones(N-1,1),1)+diag(ones(N-1,1),-1);
```

when called at the Matlab command prompt, we can now generate arbitrarily-large matrices w/ desired structure ⟶

```
>> genH(4)
ans =
    -2   1   0   0
     1  -2   1   0
     0   1  -2   1
     0   0   1  -2

>> genH(5)
ans =
    -2   1   0   0   0
     1  -2   1   0   0
     0   1  -2   1   0
     0   0   1  -2   1
     0   0   0   1  -2
```

"Diagonalization"

$\vec{X}_i$'s are orthonormal eigenvectors of Hermitian matrix $\overleftrightarrow{H}$ w/ eigenvalues λ_i.

What is $\vec{X}_i^\dagger \overleftrightarrow{H} \vec{X}_j$ ($i, j = 1, 2, \ldots$)?

Since $\overleftrightarrow{H}\vec{X}_j = \lambda_j \vec{X}_j$, $\vec{X}_i^\dagger \overleftrightarrow{H} \vec{X}_j = \vec{X}_i^\dagger \lambda_j \vec{X}_j = \lambda_j \vec{X}_i^\dagger \vec{X}_j = \lambda_j \delta_{ij} = \begin{bmatrix} \lambda_1 & 0 & \cdots \\ 0 & \lambda_2 & 0 \cdots \\ & & \ddots \end{bmatrix}$

So eigenvectors "diagonalize" the matrix!

Example

$\overleftrightarrow{H} = \begin{bmatrix} 0 & 1 \\ 1 & 0 \end{bmatrix}$ eigenvalue: $\lambda_1 = 1$ $\lambda_2 = -1$

eigvector: $\vec{X}_1 = \frac{1}{\sqrt{2}}\begin{bmatrix} 1 \\ 1 \end{bmatrix}$ $\vec{X}_2 = \frac{1}{\sqrt{2}}\begin{bmatrix} 1 \\ -1 \end{bmatrix}$

$i=1, j=1$: $X_1^\dagger H X_1 = \frac{1}{2}\begin{bmatrix} 1 & 1 \end{bmatrix}\begin{bmatrix} 0 & 1 \\ 1 & 0 \end{bmatrix}\begin{bmatrix} 1 \\ 1 \end{bmatrix} = \frac{1}{2}\begin{bmatrix} 1 & 1 \end{bmatrix}\begin{bmatrix} 1 \\ 1 \end{bmatrix} = 1$

$i=1, j=2$: $X_1^\dagger H X_2 = \frac{1}{2}\begin{bmatrix} 1 & 1 \end{bmatrix}\begin{bmatrix} 0 & 1 \\ 1 & 0 \end{bmatrix}\begin{bmatrix} 1 \\ -1 \end{bmatrix} = \frac{1}{2}\begin{bmatrix} 1 & 1 \end{bmatrix}\begin{bmatrix} -1 \\ 1 \end{bmatrix} = 0$

etc...

Matter waves: interference

EM waves: "Fabry-Perot"

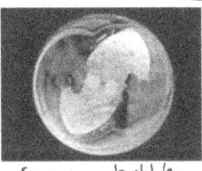
soap bubble

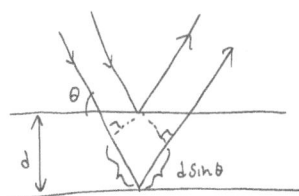

oil slick

pathlength difference = $2d\sin\theta = n\lambda$ $n = 1, 2, 3, \ldots$ for Constructive interference.

"Bragg's Law" → X-rays for $d \sim$ atomic spacing

electron waves: Davisson-Germer experiment (Nobel Prize, 1937)

De Broglie: $p = \frac{h}{\lambda}$ → $\lambda = \frac{h}{p} = \frac{h}{\sqrt{2mE}}$ $\left(E = \frac{1}{2}mv^2 = \frac{m^2v^2}{2m} = \frac{p^2}{2m}\right)$

graphite demo!

From Bragg's Law,

$d = \frac{n\lambda}{2\sin\theta} = \frac{nh}{2\sin\theta\sqrt{2mE}} \sim \frac{h}{2 \cdot 1/10\sqrt{2m \cdot 10^3 eV}} = \frac{5.4 \times 10^{-15} eV \cdot s}{\sqrt{2 \cdot 5 \times 10^{-16} eV \frac{s^2}{cm^2} \cdot 10^3 eV}} \sim 2 \text{ Å}$

atomic scale spacing! c.f. actual ~3 Å ! ✓

$E = mc^2$ → $M = E/c^2$ (mass of electron is 511 KeV/c^2)

Matter waves: interference

EM waves: Young's two-slit expt

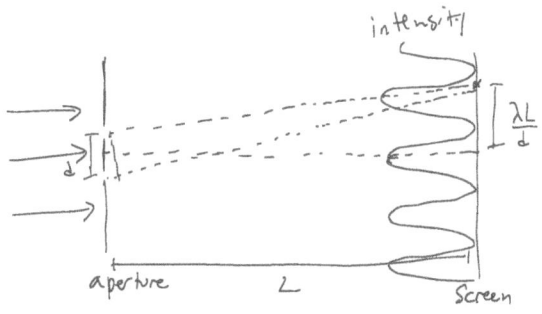

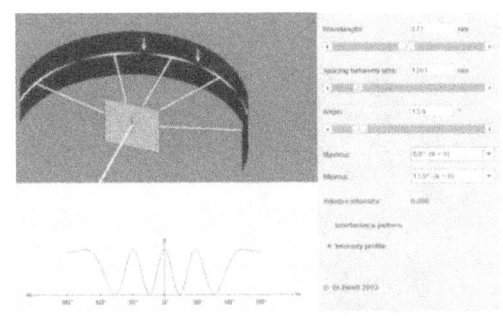

http://www.walter-fendt.de/ph14e/doubleslit.htm

electrons: Biprism interference

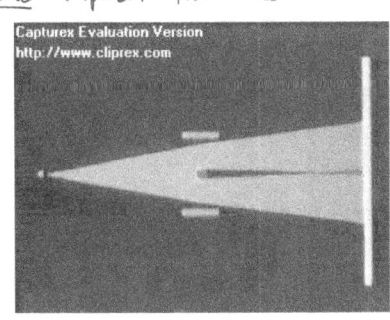

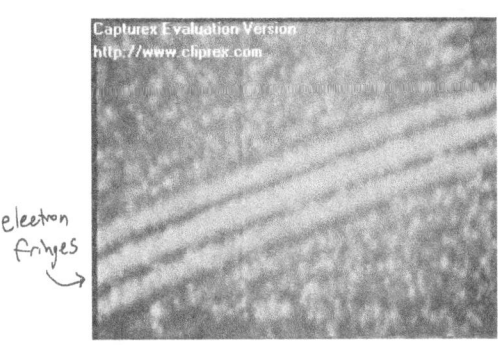

electron fringes

Electron interferometry: solid-state (semiconductors)

two-slit:

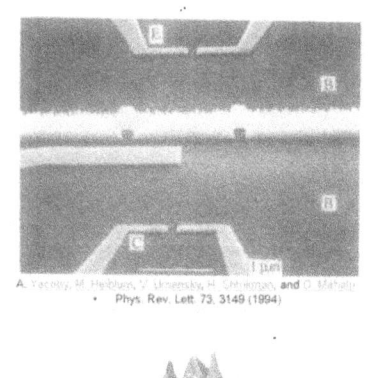

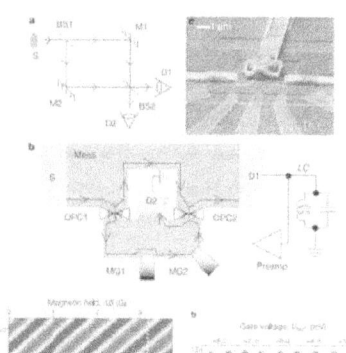

An electronic Mach–Zehnder interferometer

Matter wave equation

- What we know:

  Einstein
  $$E = hf = \left(\frac{h}{2\pi}\right)(2\pi f) = \hbar\omega$$

  de Broglie
  $$p = \frac{h}{\lambda} = \left(\frac{h}{2\pi}\right)\left(\frac{2\pi}{\lambda}\right) = \hbar k$$

- plane wave solutions $\Psi(x,t) \simeq e^{i(kx - \omega t)}$
- Correspondence w/ classical relation $E = p^2/2m + V$ (total = Kinetic + Potential)

$\Rightarrow$ Interpret $\hat{E}, \hat{p}$ (and all other observables) as linear **operators**

Measurement yields eigenvalue of appropriate operator (more on this later)

If plane wave is eigenfunction of energy and momentum,

$\hat{E}\Psi = \hbar\omega\Psi$ so $\hat{E} = i\hbar\frac{\partial}{\partial t}$ / $\hat{p}\Psi = \hbar k\Psi$ so $\hat{p} = \frac{\hbar}{i}\frac{\partial}{\partial x}$

then $\hat{E}\Psi = \left(\frac{\hat{p}^2}{2m} + \hat{V}\right)\Psi$ becomes

$$i\hbar\frac{\partial}{\partial t}\Psi(x,t) = -\frac{\hbar^2}{2m}\frac{\partial^2}{\partial x^2}\Psi(x,t) + V(x)\Psi(x,t)$$

the "time-dependent Schrödinger equation"

Simplest case: V(x)=0 ("Free particle")

Solution: plane wave $\psi(x,t) = e^{i(kx - \omega t)}$

$$-\frac{\hbar^2}{2m}\frac{\partial^2}{\partial x^2}\psi = i\hbar\frac{\partial}{\partial t}\psi$$

$$\frac{\hbar^2 k^2}{2m}\psi = \hbar\omega\psi \qquad \Rightarrow \qquad \omega = \frac{\hbar k^2}{2m} \quad \text{"dispersion relation"}$$

c.f. classical wave: $\omega = kc \quad (\lambda f = c) \quad V_{phase} = \frac{\omega}{k} = c \quad V_{group} = \frac{d\omega}{dk} = c$

For matter waves, $V_{phase} = \frac{\omega}{k} = \frac{\hbar k}{2m}$, $V_{group} = \frac{d\omega}{dk} = \frac{\hbar k}{m} = \frac{p}{m} = \frac{mv}{m} = v$

Corresponds to classical velocity!

What is the meaning of the wavefunction?

"Born Rule" / "Copenhagen Interpretation"

$|\psi|^2 = \psi^*\psi$ is a probability density

Consequences:

- Normalization: $\int_{-\infty}^{\infty} \psi^*\psi \, dx = 1$

- Probability is conserved:

  $J_L \to \quad J_R \to$ "probability current"
  $\psi^*\psi \Delta x$

  ⟵ Δx ⟶ $\to x$

$$\frac{\partial}{\partial t}(\psi^*\psi \Delta x) = J_L - J_R \quad \to \quad \frac{\partial}{\partial t}\psi^*\psi = \frac{J_L - J_R}{\Delta x}$$

In the continuous limit $\Delta x \to 0$, $\frac{\partial}{\partial t}\psi^*\psi = -\frac{\partial J}{\partial x}$ "Continuity equation"

What is probability current J?

$$\frac{\partial}{\partial t}(\psi^*\psi) = \left(\frac{\partial}{\partial t}\psi^*\right)\psi + \psi^*\left(\frac{\partial}{\partial t}\psi\right)$$

From TDSE: $\quad \frac{\partial \psi}{\partial t} = -\frac{\hbar}{2mi}\frac{\partial^2 \psi}{\partial x^2} + \frac{V}{i\hbar}\psi \quad$ and $\quad \frac{\partial \psi^*}{\partial t} = \frac{\hbar}{2mi}\frac{\partial^2 \psi^*}{\partial x^2} - \frac{V}{i\hbar}\psi^*$

$$\frac{\partial}{\partial t}(\psi^*\psi) = \frac{\hbar}{2mi}\frac{\partial^2 \psi^*}{\partial x^2}\psi - \cancel{\frac{V}{i\hbar}\psi^*\psi} - \psi^*\frac{\hbar}{2mi}\frac{\partial^2 \psi}{\partial x^2} + \cancel{\psi^*\frac{V}{i\hbar}\psi}$$

$$= -\frac{\partial}{\partial x}\left[\frac{\hbar}{2mi}\left(\psi^*\frac{\partial}{\partial x}\psi - \frac{\partial \psi^*}{\partial x}\psi\right)\right] = -\frac{\partial J}{\partial x}$$

So $\quad J = \frac{\hbar}{2mi}\left(\psi^*\frac{\partial}{\partial x}\psi - \frac{\partial \psi^*}{\partial x}\psi\right) = \text{Im}\left\{\frac{\hbar}{m}\psi^*\frac{\partial \psi}{\partial x}\right\}$

Example: free particle plane wave $\psi = e^{i(kx-\omega t)}$

$$J = \text{Im}\left\{\frac{\hbar}{m}e^{-i(kx-\omega t)}\, ik\, e^{i(kx-\omega t)}\right\} = \psi^*\psi\,\frac{\hbar k}{m} \sim \text{density} \cdot \text{velocity}.$$
$$\text{c.f. classical particle flux!}$$

Comparison with energy flux carried by E&M wave

"Poynting" energy flux: $\quad \vec{j} = \frac{1}{2}\text{Re}\{\vec{E}\times\vec{B}^*\}$

Faraday's Law: $\vec{\nabla}\times\vec{E} = -\frac{\partial \vec{B}}{\partial t} = i\omega\vec{B}\quad$ when time dependence of $\vec{B}$ is $\propto e^{-i\omega t}$ as in a propagating wave. Then, $\vec{B}^* = -\frac{1}{i\omega}\vec{\nabla}\times\vec{E}^*$

If $\vec{E} = E_z \hat{z}$ as in a wave propagating in the $\hat{x}$-$\hat{y}$ plane,

$$\vec{\nabla}\times\vec{E}^* = \begin{vmatrix} \hat{x} & \hat{y} & \hat{z} \\ \frac{\partial}{\partial x} & \frac{\partial}{\partial y} & \frac{\partial}{\partial z} \\ 0 & 0 & E_z^* \end{vmatrix} = \hat{x}\frac{\partial E_z^*}{\partial y} - \hat{y}\frac{\partial E_z^*}{\partial x} = -(\hat{z}\times\hat{y})\frac{\partial E_z^*}{\partial y} - (\hat{z}\times\hat{x})\frac{\partial E_z^*}{\partial x} = -\hat{z}\times\vec{\nabla}E_z^*$$

Therefore, $\vec{B}^* = -\frac{1}{i\omega}\vec{\nabla}\times\vec{E}^* = +\frac{1}{i\omega}\hat{z}\times\vec{\nabla}E_z^*\quad$ and

$$\vec{j} = \frac{1}{2}\text{Re}\left\{\vec{E}\times\frac{1}{i\omega}\hat{z}\times\vec{\nabla}E_z^*\right\}$$

Comparison with energy flux carried by E&M wave (cont)

Since $\vec{A} \times \vec{B} \times \vec{C} = \vec{B}(\vec{A} \cdot \vec{C}) - \vec{C}(\vec{A} \cdot \vec{B})$, this can be written

$$= \frac{1}{2\omega} \text{Re}\left\{ i \left(\hat{z} (\mathcal{E}_z \vec{\nabla} \mathcal{E}_z^*) - \vec{\nabla} \mathcal{E}_z^* (\mathcal{E}_z \cdot \hat{z}) \right) \right\}$$

The first term vanishes because of Gauss' Law:

$\hat{z} \cdot \vec{\nabla} \mathcal{E}_z^* = \frac{\partial}{\partial z} \mathcal{E}_z^*$ but since $\vec{\mathcal{E}} = \mathcal{E}_z \hat{z}$, $\vec{\nabla} \cdot \vec{\mathcal{E}}^* = \frac{\partial}{\partial z} \mathcal{E}_z^* = 0$

Then

$$\vec{j} = \frac{1}{2\omega} \text{Re}\left\{ i \mathcal{E}_z \vec{\nabla} \mathcal{E}_z^* \right\} = \frac{1}{2\omega} \text{Im}\left\{ \mathcal{E}_z^* \vec{\nabla} \mathcal{E}_z \right\} \qquad \text{(energy flux in E+M)}$$

This is mathematically identical to

$$\vec{J} = \frac{\hbar}{m} \text{Im}\left\{ \psi^* \vec{\nabla} \psi \right\} \qquad \text{(probability flux in QM)}$$

"Free particle" evolution in x and t

Time-dependent Schrödinger Eqn, $V(x)=0$: $\quad i\hbar \frac{\partial}{\partial t} \psi = -\frac{\hbar^2}{2m} \frac{\partial^2}{\partial x^2} \psi$

$$\Rightarrow \quad \frac{\partial \psi}{\partial t} = \left(\frac{\hbar i}{2m}\right) \frac{\partial^2}{\partial x^2} \psi$$

This looks just like the classical diffusion eqn. but with <u>imaginary</u> coef. $D \to iD = i\hbar/2m$!
So we know how to solve it, via separation of variables or just substitution of the Fourier form $\psi(x,t) = \frac{1}{\sqrt{2\pi}} \int_{-\infty}^{\infty} \tilde{\psi}(k,t) e^{ikx} dk$.

The result (see lec. 3) is $\psi(x,t) = \frac{1}{\sqrt{2\pi}} \int_{-\infty}^{\infty} A(k) e^{-(iD)k^2 t} e^{ikx} dk$, where $A(k)$ is the Fourier transform of initial conditions $\psi(x, t=0)$.

<u>Example</u> Initial conditions: normalized gaussian $\psi(x,t=0) = \left(\frac{2a}{\pi}\right)^{1/4} e^{-ax^2}$ (note: $\int_{-\infty}^{\infty} \psi^* \psi \, dx = 1$)

Probability density $\psi^* \psi$ has a width $\Delta x = \frac{1}{2\sqrt{a}}$.

The Fourier transform (again, see lec. 3) is $A(k) = \frac{1}{(2a\pi)^{1/4}} e^{-k^2/4a}$.

Calculation of wavefunction

Substitute $A(k)$:

$$= \left(\frac{1}{2\pi}\right)^{3/4} \frac{1}{a^{1/4}} \int_{-\infty}^{\infty} e^{-\frac{k^2}{4a}} e^{-iDk^2 t} e^{ikx} dk$$

Factor exponent:

$$= \left(\frac{1}{2\pi}\right)^{3/4} \frac{1}{a^{1/4}} \int_{-\infty}^{\infty} e^{-\left(\frac{1}{4a}+iDt\right)\left(k^2 - \frac{ikx}{\frac{1}{4a}+iDt}\right)} dk$$

Complete the square:

$$= \left(\frac{1}{2\pi}\right)^{3/4} \frac{1}{a^{1/4}} \int_{-\infty}^{\infty} e^{-\left(\frac{1}{4a}+iDt\right)\left(k - \frac{ix}{2\left(\frac{1}{4a}+iDt\right)}\right)^2} e^{-\frac{x^2}{4\left(\frac{1}{4a}+iDt\right)}} dk$$

Perform definite integral over gaussian:

$$= \left(\frac{1}{2\pi}\right)^{3/4} \frac{1}{a^{1/4}} \sqrt{\frac{\pi\, 4a}{1+4iDat}} \; e^{-\frac{ax^2}{1+4iDat}} = \left(\frac{2a}{\pi}\right)^{1/4} \frac{1}{\sqrt{1+4iDat}} \; e^{-\frac{ax^2}{1+4iDat}}$$

Don't panic about a complex gaussian! Only $\psi^*\psi$ is physically meaningful!

Probability density evolution

$$\psi^*\psi \propto e^{-\frac{ax^2}{1-4iDat}} e^{-\frac{ax^2}{1+4iDat}} = e^{-2ax^2\left(\frac{1}{1+16D^2 a^2 t^2}\right)} = e^{-\frac{x^2}{2\Delta x^2}} \text{ where } \Delta x = \sqrt{\frac{1+16D^2 a^2 t^2}{4a}},\; D = \frac{\hbar}{2m}$$

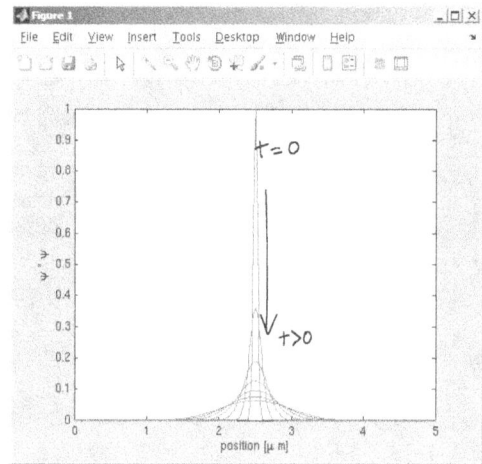
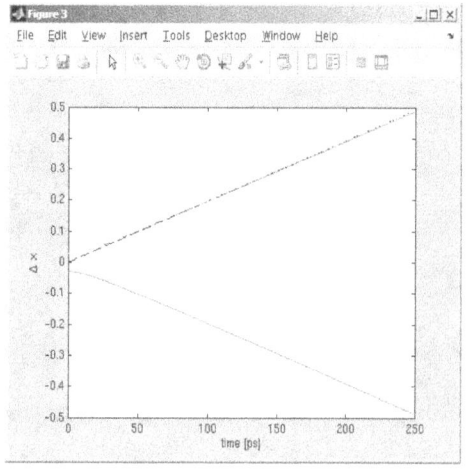

Just like the classical diffusion equation, the distribution widens and the maximum decreases to conserve the total probability $\int_{-\infty}^{\infty} \psi^*\psi\, dx = 1$

C.f. sol'n to classical equation of motion for $V(x)=0$, and I.C.s $x(t=0)=0, \dot{x}(t=0)=0$!

Heisenberg uncertainty principle

Note that for large t when $D^2 a^2 t^2 \gg 1$, $\Delta x \approx 2D\sqrt{a}\, t$. This means that the gaussian grows **Faster** for more localized initial conditions (Δx small / $\sqrt{a}$ large). This is the "Heisenberg" uncertainty principle $\Delta x \Delta p \geq \hbar/2$ at work: small Δx gives large momentum uncertainty Δp, which causes faster broadening!

An analogy:
This looks like an optical beam emerging from a focal point along its propagation axis!
$\Rightarrow$ The <u>uncertainty principle</u> is also the reason why you should use the shortest focal distance lens w/ the largest aperture diameter to get the smallest focal point and highest optical resolution! (microscopes / telescopes)

Derivation of "optical Schrödinger equation"

Classical wave equation governing E+M wave propagation:
$$\nabla^2 \vec{\mathcal{E}} = \frac{1}{c^2} \frac{\partial^2}{\partial t^2} \vec{\mathcal{E}}$$

Assume the beam is cylindrically-symmetric about propagation direction $\hat{z}$ and polarized along $\hat{x}$:
$$\vec{\mathcal{E}} = \mathcal{E}_x(x,z)\, e^{i(kz - \omega t)} \hat{x}$$

Then,
$$\frac{\partial^2 \mathcal{E}_x}{\partial x^2} e^{i(kz-\omega t)} + \frac{\partial^2}{\partial z^2}\left(\mathcal{E}_x e^{i(kz-\omega t)}\right) = \frac{1}{c^2} \frac{\partial^2}{\partial t^2} \mathcal{E}_x e^{i(kz-\omega t)}$$

Distributing the derivative and cancelling common terms yields:
$$\frac{\partial^2 \mathcal{E}_x}{\partial x^2} e^{i(kz-\omega t)} + \frac{\partial}{\partial z}\left(\frac{\partial \mathcal{E}_x}{\partial z} e^{i(kz-\omega t)} + ik \mathcal{E}_x e^{i(kz-\omega t)}\right) = \frac{-\omega^2}{c^2} \mathcal{E}_x e^{i(kz-\omega t)}$$

$$\frac{\partial^2 \mathcal{E}_x}{\partial x^2} e^{i(kz-\omega t)} + \frac{\partial^2}{\partial z^2}\mathcal{E}_x e^{i(kz-\omega t)} + ik \frac{\partial \mathcal{E}_x}{\partial z} e^{i(kz-\omega t)} + ik\left(\frac{\partial}{\partial z}\mathcal{E}_x e^{i(kz-\omega t)} + ik \mathcal{E}_x e^{i(kz-\omega t)}\right) = \frac{-\omega^2}{c^2} \mathcal{E}_x e^{i(kz-\omega t)}$$

$$\frac{\partial^2 \mathcal{E}_x}{\partial x^2} + \frac{\partial^2}{\partial z^2}\mathcal{E}_x + 2ik \frac{\partial \mathcal{E}_x}{\partial z} - k^2 \mathcal{E}_x = \frac{-\omega^2}{c^2} \mathcal{E}_x$$

Slowly-varying envelope approximation

Now, we know that $\omega = kc$ (classical wave dispersion), so:

$$\frac{\partial^2 \mathcal{E}_x}{\partial x^2} + \frac{\partial^2}{\partial z^2}\mathcal{E}_x + 2ik\frac{\partial \mathcal{E}_x}{\partial z} = 0$$

Notice that if we ignore $\frac{\partial^2}{\partial z^2}\mathcal{E}_x$ ("Slowly varying envelope approx"), we have

$$\frac{\partial^2 \mathcal{E}_x}{\partial x^2} = (-2k)i\frac{\partial \mathcal{E}_x}{\partial z}$$

This is mathematically the same as our Schrödinger equation

$$\frac{\partial^2 \psi}{\partial x^2} = \left(\frac{2m}{\hbar}\right)i\frac{\partial \psi}{\partial t}$$

but w/ different coef. and $t \to z$! So, if the beam has a gaussian intensity cross-section, we can expect it to diffract along propagation direction in approx. same way as QM probability evolves in time!

Expectation values of observables

In QM, every observable has a corresponding operator.
For any observable Q, the possible outcome of measurement are the eigenvalues of the corresponding operator $\hat{Q}$

$$\hat{Q}X_i = q_i X_i \qquad (X_i \text{ are eigenfunctions/eigenvectors and } q_i \text{ are corresponding eigenvalues})$$

But, in general, ψ is not an eigenfunction. But, $\psi = \sum_i b_i X_i$ so

$$\hat{Q}\psi = \sum_i b_i q_i X_i \qquad \text{Now take inner product w/ wavefunction}$$

$$\int_{-\infty}^{\infty} \psi^* \hat{Q} \psi \, dx = \int_{-\infty}^{\infty} \psi^* \sum_i b_i q_i X_i \, dx = \int_{-\infty}^{\infty} \sum_j b_j^* X_j^* \sum_i b_i q_i X_i \, dx$$

$$= \sum_i \sum_j b_j^* b_i q_i \int_{-\infty}^{\infty} X_j^* X_i \, dx = \sum_i \sum_j b_j^* b_i q_i \delta_{ij} \quad \underbrace{}_{\text{orthonormality}}$$

$$= \sum_i |b_i|^2 q_i = \langle Q \rangle \qquad \text{"expectation value" is a weighted sum of all possibilities!}$$

So $\langle Q \rangle = \int_{-\infty}^{\infty} \psi^* \hat{Q} \psi \, dx$

Examples associated with mechanical motion

Position $\quad \langle x \rangle = \int_{-\infty}^{\infty} \psi^* x \psi \, dx = \int_{-\infty}^{\infty} x \, \psi^* \psi \, dx$

Since $\psi^* \psi$ is probability density, this is the same as classical distribution theory!

e.g. mass density $\rho(x)$, $0 \to L \to x$, center of mass $\langle x \rangle = \dfrac{\int_0^L x \rho(x) dx}{\int_0^L \rho(x) dx}$ (Note that $\int_{-\infty}^{\infty} \psi^* \psi \, dx = 1$)

velocity $\langle v \rangle = \dfrac{d\langle x \rangle}{dt} = \int_{-\infty}^{\infty} x \dfrac{d}{dt}(\psi^* \psi) dx \xrightarrow{\text{continuity}} \int_{-\infty}^{\infty} x \left(-\dfrac{dJ}{dx}\right) dx$

(probability density, probability current)

$= -J(x) x \Big|_{-\infty}^{\infty}{}^{\to 0} + \int_{-\infty}^{\infty} J \, dx \quad$ where $J = \dfrac{\hbar}{2mi}\left(\psi^* \dfrac{d}{dx}\psi - \dfrac{d}{dx}\psi^* \cdot \psi\right)$

Since ψ must be normalized, $\psi(x = \pm\infty) \to 0$, so $J(x = \pm\infty) = 0$

$= \dfrac{\hbar}{2mi}\left[\int_{-\infty}^{\infty} \psi^* \dfrac{d}{dx}\psi \, dx - \int_{-\infty}^{\infty} \dfrac{d}{dx}\psi^* \cdot \psi \, dx\right]$

Examples associated with mechanical motion (cont)

Integration by parts on second term yields:

$-\int_{-\infty}^{\infty} \dfrac{d}{dx}\psi^* \cdot \psi \, dx = -\psi^* \psi \Big|_{-\infty}^{\infty}{}^{\to 0} + \int_{-\infty}^{\infty} \psi^* \dfrac{d}{dx}\psi \, dx \Rightarrow$ The first term!

So $\quad \langle v \rangle = \dfrac{\hbar}{mi} \int_{-\infty}^{\infty} \psi^* \dfrac{d}{dx}\psi \, dx$

Therefore

$\langle p \rangle = m \langle v \rangle = \int_{-\infty}^{\infty} \psi^* \dfrac{\hbar}{i} \dfrac{d}{dx} \psi \, dx$

As expected, since $\hat{p} = \dfrac{\hbar}{i}\dfrac{d}{dx}$!

We could have just used $\int_{-\infty}^{\infty} \psi^* \hat{Q} \psi \, dx$!

Uncertainty from expectation values

The standard deviation of observable Q is given by

$$\sigma_Q = (\text{Variance})^{1/2} = \left[\int_{-\infty}^{\infty} \psi^* (Q - \langle Q \rangle)^2 \psi \, dx \right]^{1/2}$$

$$= \left[\int_{-\infty}^{\infty} \psi^* (Q^2 - 2Q\langle Q \rangle + \langle Q \rangle^2) \psi \, dx \right]^{1/2}$$

$$= \left[\int_{-\infty}^{\infty} \psi^* Q^2 \psi \, dx - 2\langle Q \rangle \int_{-\infty}^{\infty} \psi^* Q \psi \, dx + \langle Q \rangle^2 \int \psi^* \psi \, dx \right]^{1/2}$$

$$= \left(\langle Q^2 \rangle - 2\langle Q \rangle^2 + \langle Q \rangle^2\right)^{1/2} = \left(\langle Q^2 \rangle - \langle Q \rangle^2\right)^{1/2}$$

For gaussian state above, $\sigma_x = \sqrt{\langle x^2 \rangle - \cancel{\langle x \rangle^2}^{\,0}}$ by symmetry
$\sigma_p = \sqrt{\langle p^2 \rangle - \cancel{\langle p \rangle^2}^{\,0}}$

Direct calculation of Δx   (free particle wavefunction)

$$\psi(x,t) = \left(\frac{2a}{\pi}\right)^{1/4} \frac{1}{\sqrt{1+4iDat}} e^{-\frac{ax^2}{1+4iDat}} \qquad \left(D = \frac{\hbar}{2m}\right)$$

$$\text{"}\Delta x\text{"} = \sigma_x = \sqrt{\langle x^2 \rangle - \cancel{\langle x \rangle^2}^{\,0}}$$

$$\langle x^2 \rangle = \int_{-\infty}^{\infty} \psi^* x^2 \psi \, dx = \sqrt{\frac{2a}{\pi(1+16D^2a^2t^2)}} \int_{-\infty}^{\infty} x^2 e^{-\frac{2ax^2}{1+16D^2a^2t^2}} dx$$

$$= \sqrt{\frac{\xi}{\pi}} \int_{-\infty}^{\infty} x^2 e^{-\xi x^2} dx \qquad \xi = \frac{2a}{1+16D^2a^2t^2}$$

$$(\text{differentiate under integral}) = \sqrt{\frac{\xi}{\pi}} \int_{-\infty}^{\infty} \left(-\frac{d}{d\xi}\right) e^{-\xi x^2} dx = -\sqrt{\frac{\xi}{\pi}} \frac{d}{d\xi} \left[\int_{-\infty}^{\infty} e^{-\xi x^2} dx\right]$$

$$= -\sqrt{\frac{\xi}{\pi}} \frac{d}{d\xi} \sqrt{\frac{\pi}{\xi}} = \sqrt{\frac{\xi}{\pi}} \frac{1}{2} \frac{\sqrt{\pi}}{\xi^{3/2}} = \frac{1}{2\xi} = \frac{1+16D^2a^2t^2}{4a}$$

So $\quad \text{"}\Delta x\text{"} = \sigma_x = \sqrt{\langle x^2 \rangle} = \sqrt{\frac{1+16D^2a^2t^2}{4a}}$

Verifying the Heisenberg uncertainty principle

$"\Delta p" = \sigma_p = \sqrt{\langle p^2 \rangle - \cancel{\langle p \rangle^2}^{\;0 \text{ by symmetry}}}$ $\Psi(x,t) = \left(\frac{2a}{\pi}\right)^{1/4} \frac{1}{\sqrt{1+4iDat}} e^{-\frac{ax^2}{1+4iDat}}$

$\langle p^2 \rangle = \int_{-\infty}^{\infty} \Psi^* \left(\frac{\hbar}{i}\frac{d}{dx}\right)^2 \Psi \, dx = -\hbar^2 \int_{-\infty}^{\infty} \Psi^* \frac{d}{dx}\left(-\frac{2ax}{1+4iDat}\Psi\right) dx$

$= \frac{2a\hbar^2}{1+4iDat} \int_{-\infty}^{\infty} \Psi^* \left(\Psi + x\left(-\frac{2ax}{1+4iDat}\Psi\right)\right) dx = \frac{2a\hbar^2}{1+4iDat}\left[\int_{-\infty}^{\infty}\cancelto{1}{\Psi^*\Psi}\,dx - \frac{2a}{1+4iDat}\cancelto{\langle x^2 \rangle}{\int_{-\infty}^{\infty}\Psi^* x^2 \Psi\, dx}\right]$

$= \frac{2a\hbar^2}{1+4iDat}\left[1 - \frac{2a}{1+4iDat}\cdot\frac{(1+4iDat)(1-4iDat)}{4a}\right] = \frac{2a\hbar^2}{\cancel{1+4iDat}}\cdot\frac{1}{2}(\cancel{1+4iDat}) = a\hbar^2$

So $"\Delta p" = \sigma_p = \sqrt{\langle p^2 \rangle} = \hbar\sqrt{a}$ time independent!

Spatial translation invariance → Momentum is conserved

$"\Delta x""\Delta p" = \sigma_x \sigma_p = \sqrt{\frac{1+16D^2a^2t^2}{4a}}\,\hbar\sqrt{a} = \frac{\hbar}{2}\cdot(1+16D^2a^2t^2)^{1/2} \geq \frac{\hbar}{2}$ ✓

General case (V(x)≠0)

$$-\frac{\hbar^2}{2m}\frac{\partial^2}{\partial x^2}\Psi(x,t) + V(x)\Psi(x,t) = i\hbar\frac{\partial}{\partial t}\Psi(x,t)$$

Separation of variables: $\Psi(x,t) = \varphi(x)\phi(t)$

$$\frac{-\frac{\hbar^2}{2m}\varphi''\phi + V\varphi\phi}{\varphi\phi} = \frac{i\hbar\varphi\dot\phi}{\varphi\phi}$$

$$\frac{-\frac{\hbar^2}{2m}\varphi'' + V\varphi}{\varphi} = \frac{i\hbar\dot\phi}{\phi} = E \qquad \text{only if } V(x,t) \to V(x)$$

time invariance → energy conservation

For x:

$-\frac{\hbar^2}{2m}\varphi'' + V(x)\varphi = E\varphi$

"Time-independent Schrödinger Eqn"

For t:

$\dot\phi = -i\frac{E}{\hbar}\phi \quad \to \quad \phi(t) = Ce^{-i\frac{E}{\hbar}t}$

In "stationary" potential, time dependence of $\Psi(x,t)$ is just an overall phase!

Infinite square well

$$V(x) = \begin{cases} 0 & 0 < x < a \\ \infty & \text{otherwise} \end{cases}$$

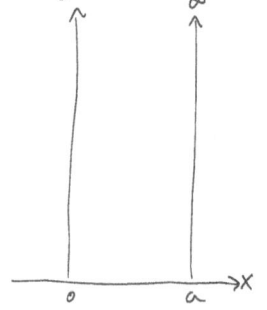

$$\varphi'' = -\frac{2m(E-V)}{\hbar^2}\varphi = -k^2\varphi$$

Solutions: $\varphi(x) = Ae^{ikx} + Be^{-ikx}$, $k \equiv \sqrt{\frac{2m(E-V)}{\hbar^2}}$

In regions $x<0$ and $x>a$, $k = \pm i\text{``}\infty\text{''}$ so $\varphi \to \infty$ unless $A, B = 0$

This means we have B.C.'s: $\varphi(0) = 0$ and $\varphi(a) = 0$

$\varphi(x) = A'\sin kx + B'\cos kx$ $\quad (0 < x < a)$

Impose $\varphi(0) = B' = 0$

$\quad \varphi(a) = A'\sin ka = 0 \longrightarrow k_n = \frac{n\pi}{a}$ $\quad n = 1, 2, 3, \ldots$ "quantum number"

From k_n, we determine the Energy Eigenvalue $\longrightarrow$ $E = \frac{\hbar^2 k^2}{2m} = \frac{\hbar^2 n^2 \pi^2}{2ma^2}$

Normalization

$$\int_{-\infty}^{\infty} \psi^* \psi \, dx = 1$$

$$\int_0^a A'^2 \sin^2 kx \, dx = A'^2 \int_0^a \frac{1 - \cos 2kx}{2} dx = A'^2 \left[\frac{x}{2} - \frac{\sin 2kx}{4k}\right]\Big|_0^a = A'^2 \frac{a}{2} = 1$$

So $A' = \sqrt{\frac{2}{a}}$

eigenfunctions and associated eigenvalues are

$$\varphi_n(x) = \sqrt{\frac{2}{a}} \sin \frac{n\pi x}{a}, \quad \Psi_n(x,t) = \begin{cases} \sqrt{\frac{2}{a}} \sin \frac{n\pi x}{a} e^{-\frac{iE_n t}{\hbar}}, & 0 < x < a, \; E_n = \frac{\hbar^2 \pi^2 n^2}{2ma^2} \\ 0 & \text{otherwise} \end{cases}$$

Remember: Any wavefunction $\Psi(x,t)$ can be decomposed into a linear superposition of eigenfunctions $\Psi_n(x,t)$

Finite square well

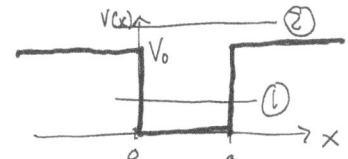

In each piecewise const. region,
$$\psi'' = -K^2 \psi$$
$$K = \sqrt{\frac{2m(E-V)}{\hbar^2}}, \quad V=0 \text{ or } V_0$$

2 types of states:

1) $0 < E < V_0$ "bound" states

$$\psi(x) \to \begin{cases} A_+ e^{+iK_1 x} + A_- e^{-iK_1 x}, & 0 < x < a \quad \left(K_1 = \sqrt{\frac{2mE}{\hbar^2}}\right) \\ B_+ e^{+iK_2 x} + B_- e^{-iK_2 x}, & x < 0 \text{ or } x > a \quad \left(K_2 = \sqrt{\frac{2m(E-V_0)}{\hbar^2}} \to \text{imaginary!}\right) \end{cases}$$

So bound states <u>decay</u> as $B_+ e^{+K_2 x}$ or $B_- e^{-K_2 x}$ $\left(K_2 = \sqrt{\frac{2m(V_0-E)}{\hbar^2}}\right)$ in "classically forbidden" region where $E < V_0$. (Only one coef is nonzero to maintain normalizable ψ)

2) $E > V_0$ "continuum" or "scattering" states

$$\psi(x) \to \begin{cases} A'_+ e^{iK_1 x} + A'_- e^{-iK_1 x}, & 0 < x < a, \quad K_1 = \sqrt{\frac{2mE}{\hbar^2}} \\ B'_+ e^{iK_2 x} + B'_- e^{-iK_2 x}, & x < 0 \text{ or } x > a \quad K_2 = \sqrt{\frac{2m(E-V_0)}{\hbar^2}} \end{cases}$$

How to determine A's and B's to "stitch" wavefunction together?
$\Rightarrow$ We need Boundary conditions!

Boundary Conditions

① $\psi(x)$ must be continuous at boundaries so $\frac{d\psi}{dx}$ is finite. Then,
$J = \text{Im}\left\{\frac{\hbar}{m} \psi^* \frac{d\psi}{dx}\right\}$ is finite + continuity holds.

② $-\frac{\hbar^2}{2m} \frac{d^2}{dx^2} \psi = (E-V)\psi$

Suppose a boundary exists @ $x=0$. Integrate from one side to the other:
$$-\frac{\hbar^2}{2m} \int_{-\epsilon}^{\epsilon} \frac{d^2}{dx^2} \psi \, dx = \int_{-\epsilon}^{\epsilon} (E-V)\psi \, dx \xrightarrow{\lim_{\epsilon \to 0}} 0 \quad \text{for finite } V(x)$$

By fundamental Thm of Calculus, we then have:
$$\frac{d}{dx}\psi\bigg|_{x=\epsilon} - \frac{d}{dx}\psi\bigg|_{x=-\epsilon} = 0 \quad \text{so} \quad \frac{d}{dx}\psi \equiv \psi' \text{ is continuous at a boundary!}$$

With these 2 Boundary conditions per N interfaces, we need to solve a large system of $2N$ equations $\to$ transcendental equations! We can avoid this problem by numerically solving eigenvalues + eigenvectors!

Numerical solution of the time-independent Schrödinger equation

$$\underbrace{\left[-\frac{\hbar^2}{2m}\frac{d^2}{dx^2}+V(x)\right]}_{\text{"Hamiltonian"}}\psi(x)=E\psi(x) \implies \hat{H}\psi=E\psi$$

Transform differential $\hat{H}$ into matrix operator:

$$-\frac{\hbar^2}{2m}\frac{d^2}{dx^2}\psi(x) \longrightarrow \frac{-\hbar^2}{2m\Delta x^2}\begin{bmatrix}-2 & 1 & 0 & \cdots \\ 1 & -2 & 1 & \\ 0 & 1 & -2 & 1 \\ \vdots & & 1 & -2 & 1 \\ & & & & \ddots\end{bmatrix}\begin{bmatrix}\psi(x_1)\\ \psi(x_2)\\ \vdots \end{bmatrix}$$

(just like classical operator!)

$$V(x)\psi(x) \longrightarrow \begin{bmatrix}V(x_1) & 0 & \cdots \\ 0 & V(x_2) & \\ \vdots & & \ddots\end{bmatrix}\begin{bmatrix}\psi(x_1)\\ \psi(x_2)\\ \vdots\end{bmatrix}$$

In actual calculations, I suggest the following system of units:

$\hbar \sim 6.6\times 10^{-16}$ eV·s

$m \sim 5\times 10^5$ eV/c^2

$c \sim 3\times 10^{10}$ cm/s

$\Delta x \sim 10^{-9}$ cm

$N \sim$ several hundred (matrix size)

$\left[\frac{\hbar^2}{2m\Delta x^2}\right] = \frac{eV^{\cancel{2}}\cdot \cancel{s^2}}{\cancel{eV}\frac{\cancel{s^2}}{\cancel{cm^2}}\cancel{cm^2}} = eV!$

Results: 1nm-wide infinite QW

100×100 matrix Hamiltonian

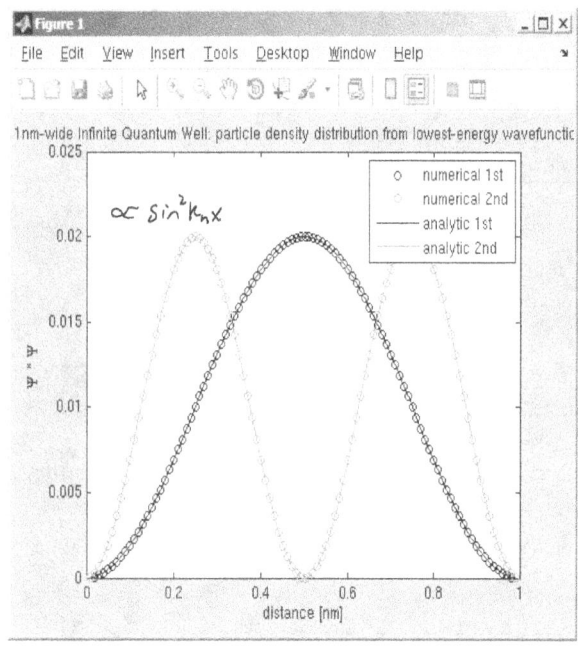
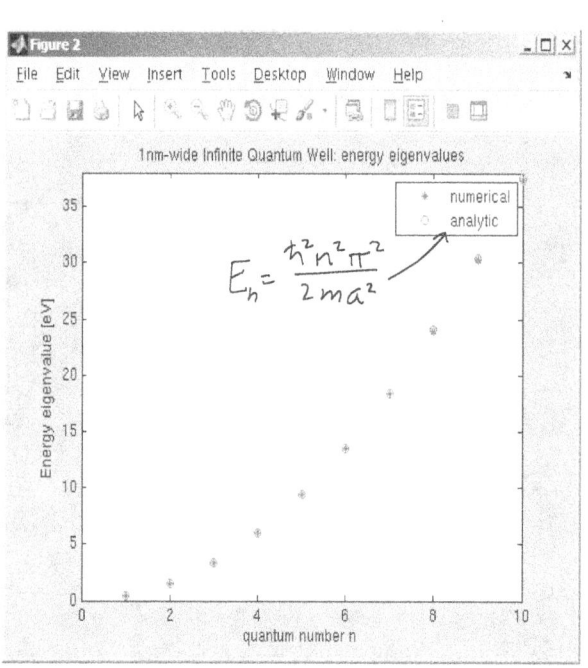

$\propto \sin^2 k_n x$

$E_n = \frac{\hbar^2 n^2 \pi^2}{2ma^2}$

Results: 1nm-wide, 1eV-deep finite QW

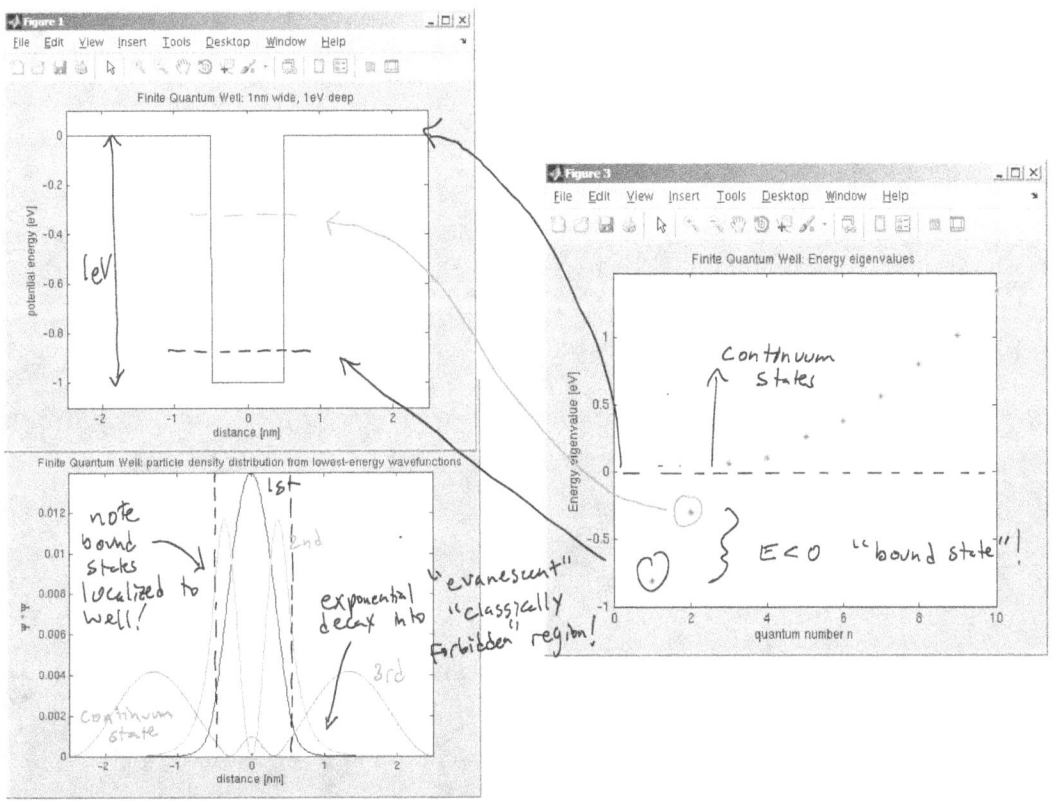

Two finite QWs, coupled by wide barrier

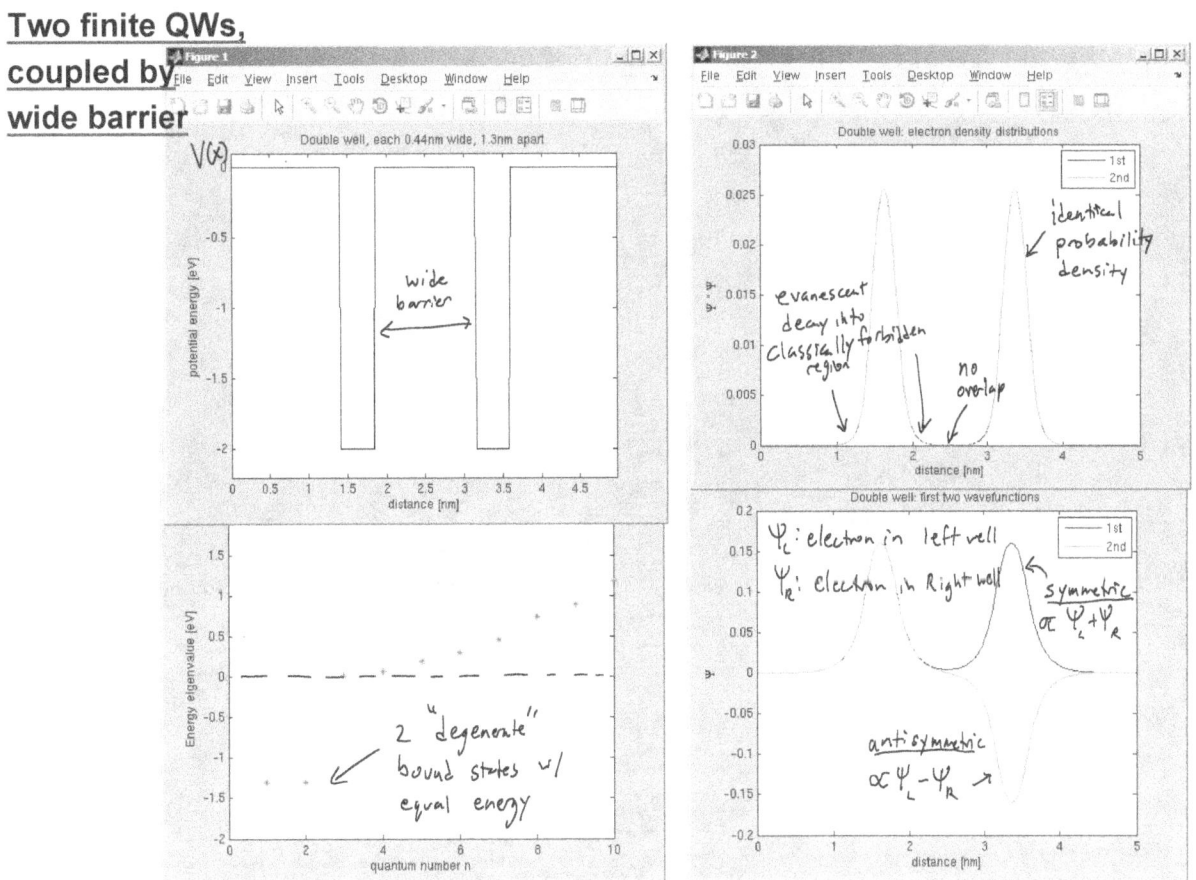

Two finite QWs, coupled by thin barrier

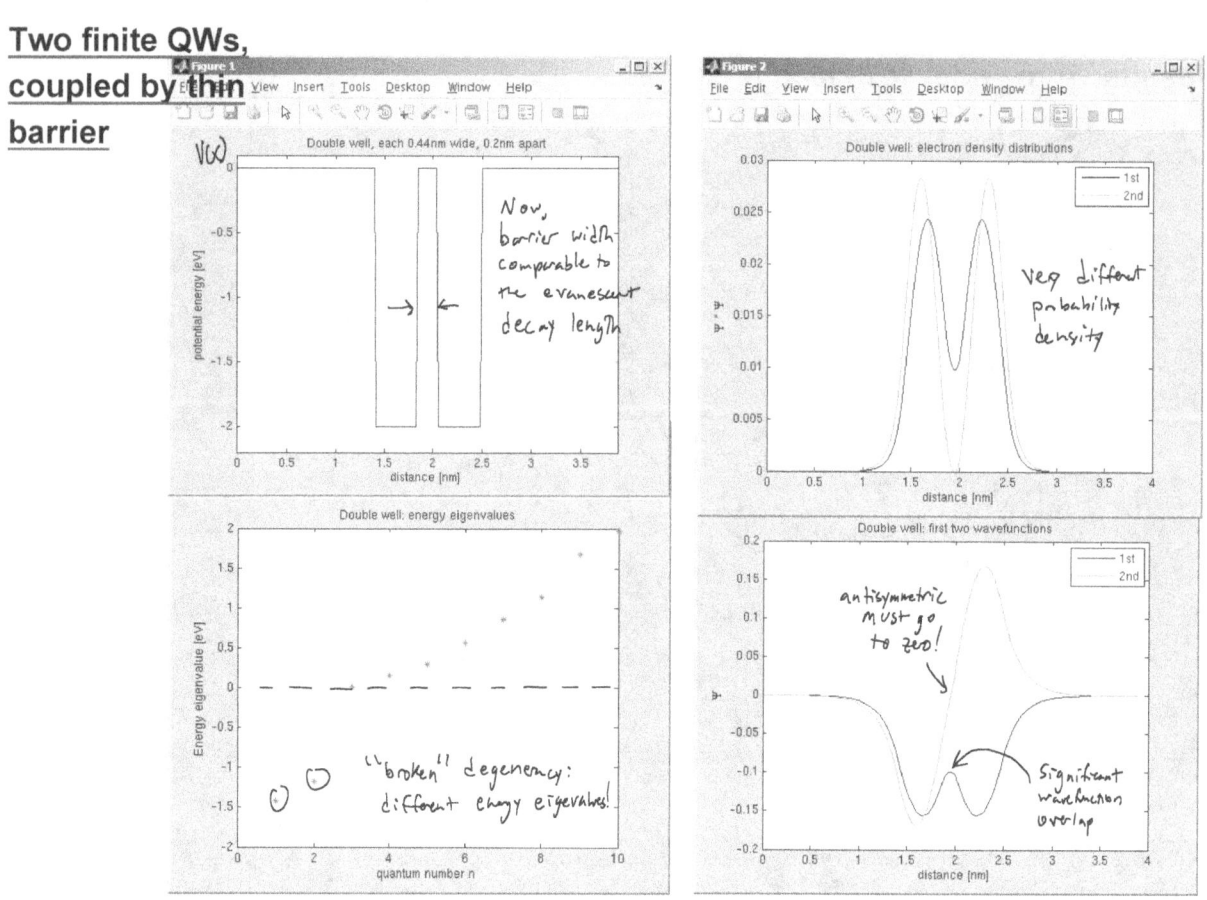

Coupled finite QWs: bound state "repulsion"

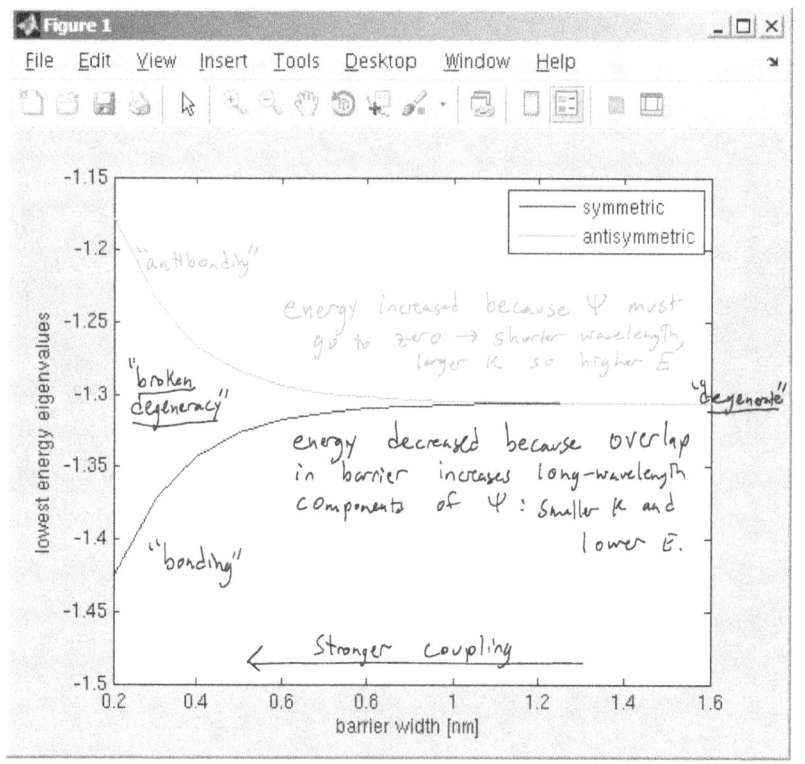

Exception to derivative continuity boundary condition

Example: $V(x) = -\alpha\delta(x)$ $\quad\left(\int\delta(x)dx = 1\right)$

$$-\frac{\hbar^2}{2m}\varphi'' - \alpha\delta(x)\varphi = E\varphi$$

$$-\frac{\hbar^2}{2m}\int_{-\epsilon}^{\epsilon}\varphi'' dx - \alpha\int_{-\epsilon}^{\epsilon}\delta(x)\varphi\, dx = \int_{-\epsilon}^{\epsilon} E\varphi\, dx$$

$\lim\limits_{\epsilon\to 0}:\quad -\frac{\hbar^2}{2m}\left(\varphi'\big|_{x=\epsilon} - \varphi'\big|_{x=-\epsilon}\right) - \alpha\varphi(0) = 0$

$$\varphi'\big|_{x=\epsilon} - \varphi'\big|_{x=-\epsilon} = -\frac{2m\alpha}{\hbar^2}\varphi(0) \qquad \underline{NOT}\text{ zero!}$$

<u>Note:</u> Since $\int_{-\infty}^{\infty}\delta(x)\,dx = 1$, since dx has units of length, $\delta(x)$ must have units of length^{-1}. Therefore, since $V(x)$ has units of energy, α has units of Energy·length.

Bound state of attractive delta function potential

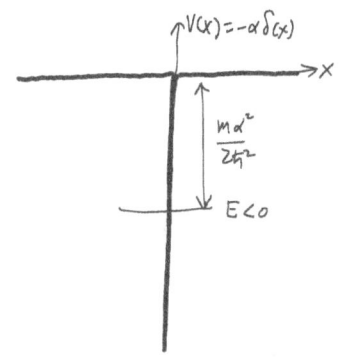

$$-\frac{\hbar^2}{2m}\varphi'' = E\varphi \qquad (x \neq 0)$$

$x < 0:\quad \varphi_-(x) = Ae^{\kappa x} + \cancel{B}e^{-\kappa x}, \quad \kappa = \sqrt{\frac{-2mE}{\hbar^2}}$

$x > 0:\quad \varphi_+(x) = \cancel{A'}e^{\kappa x} + B'e^{-\kappa x}$

B.C.'s: $\varphi \to 0$ as $x \to \pm\infty$ so that it is normalizable $(A' = B = 0)$

φ is continuous across $x = 0$ $\quad (A = B')$

$\varphi'_+(0) - \varphi'_-(0) = -\frac{2m\alpha}{\hbar^2}\varphi(0)$

$\varphi'_+(0) - \varphi'_-(0) = -\kappa B' - \kappa A = -2\kappa A = -\frac{2m\alpha}{\hbar^2}A \quad\longrightarrow\quad \kappa = \frac{m\alpha}{\hbar^2}$

$E = -\frac{\hbar^2\kappa^2}{2m} = -\frac{m\alpha^2}{2\hbar^2}$

Normalization of delta function bound state

$$\psi_-(x) = Ae^{\kappa x}, \quad \psi_+(x) = Ae^{-\kappa x}$$

$$\int_{-\infty}^{\infty} \psi^*\psi \, dx = \int_{-\infty}^{0} \psi_-^* \psi_- \, dx + \int_{0}^{+\infty} \psi_+^* \psi_+ \, dx = 2\int_{0}^{\infty} A^2 e^{-2\kappa x} dx = -\frac{2A^2}{2\kappa} e^{-2\kappa x}\Big|_0^\infty = \frac{A^2}{\kappa} = 1$$

So $A = \sqrt{\kappa} = \frac{\sqrt{m\alpha}}{\hbar}$

$$\psi(x) = \frac{\sqrt{m\alpha}}{\hbar} e^{-\frac{m\alpha}{\hbar^2}|x|}$$

Numerical results

Since $-\alpha\delta(x)$ is the analytic limit of a sequence of narrower, deeper wells with constant $\alpha = \int_0^{\text{width}} V(x) dx$, we can use our matrix eigenvalue scheme to numerically approximate the solution:

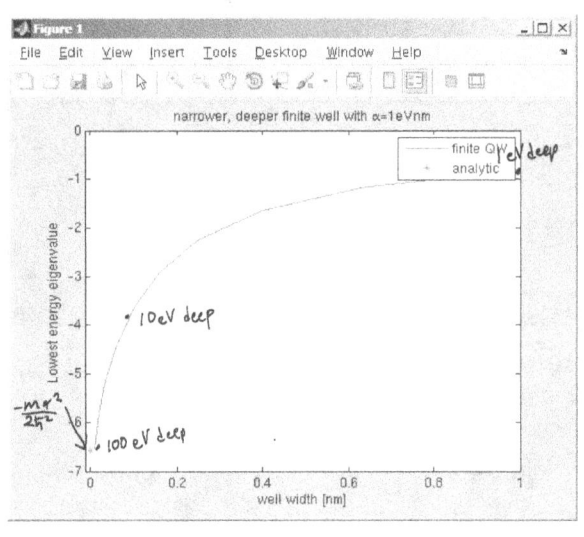
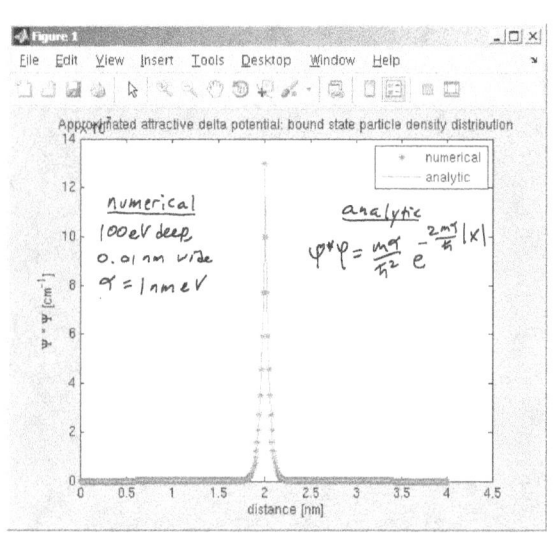

Periodic potentials (a simple model for crystalline solid)

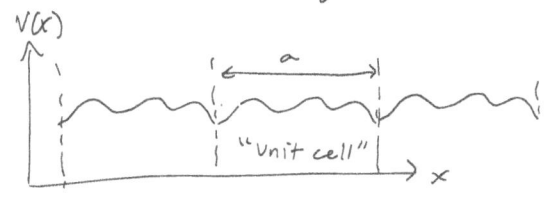

$V(x) = V(x+a)$ a is "lattice constant"

$= \sum_g c(g) e^{igx}$ $g = 0, \pm\frac{2\pi}{a}, \pm\frac{4\pi}{a}, \ldots$

(Fourier series)   is "reciprocal lattice number"

Our solution has the same symmetry as the potential:

$\psi(x) = e^{ikx} u(x)$ "Bloch wave"   where $u(x) = \sum_g b(g) e^{igx}$

__Schrödinger Eq:__

$$-\frac{\hbar^2}{2m} \psi'' + V\psi = E\psi$$

substitute:

$$-\frac{\hbar^2}{2m}\left(e^{ikx} u(x)\right)'' + V e^{ikx} u(x) = E e^{ikx} u(x)$$

Now, since $(f_1 f_2)'' = (f_1' f_2 + f_1 f_2')' = f_1'' f_2 + 2 f_1' f_2' + f_1 f_2''$,

$$-\frac{\hbar^2}{2m}\left[-k^2 e^{ikx} u(x) + 2ik e^{ikx} u'(x) + e^{ikx} u''(x)\right] + V e^{ikx} u(x) = E e^{ikx} u(x)$$

Periodic potentials (cont)

$$-\frac{\hbar^2}{2m}\left(u'' + 2ik u' - k^2 u\right) + Vu = Eu$$

$$-\frac{\hbar^2}{2m}\left(\frac{d}{dx} + ik\right)^2 u + Vu = Eu$$

because: $u(x) = \sum_g b(g) e^{igx}$ $V(x) = \sum_g c(g) e^{igx}$ so $\frac{d}{dx} \to ig$:

$$-\frac{\hbar^2}{2m} \sum_g (ik + ig)^2 b(g) e^{igx} + \sum_{g'} c(g') e^{ig'x} \sum_g b(g) e^{igx} = E \sum_g b(g) e^{igx}$$

(g' has same discrete values as g but is summed over separately)

$$\frac{\hbar^2}{2m} \sum_g (k+g)^2 b(g) e^{igx} + \sum_{g'} c(g') e^{ig'x} \sum_g b(g) e^{igx} = E \sum_g b(g) e^{igx}$$

This is __hard__ to solve for $b(g)$ exactly if the sums are __infinite__ so we have to terminate them to calculate. But first, look at the simplest case...

When V(x)→0 [all c(g)=0]

Then we retain only the periodicity of the potential:

$$\frac{\hbar^2}{2m}\sum_g (K+g)^2 b(g) e^{igx} = E \sum_g b(g) e^{igx}$$

because of orthonormality of basis fns

$$E = \frac{\hbar^2}{2m}(K+g)^2:$$

[reminder: K is continuous, g is discrete $(=0, \pm\frac{2\pi}{a}, \pm\frac{4\pi}{a}, ...)$]

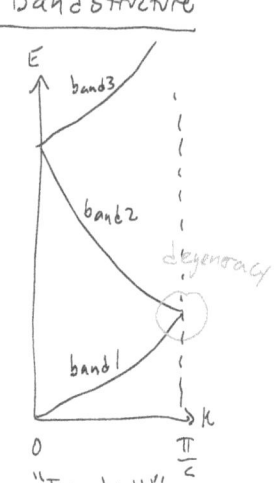

"Band structure"

"Irreducible BZ"

Like the free-particle dispersion only repeated periodically!

When V(x)≠0

Consider the degeneracy (circled in previous slide) at $K=\frac{\pi}{a}$. Then $\lambda = \frac{2\pi}{K} = 2a$:

ψ_1 and ψ_2 have the same wavelength, but are orthogonal as required for eigenstates.

Probability density $\psi^*\psi$ shows that an electron in eigenstate ψ_1 is more likely to be found at regions where V(x) is large, and ψ_2 is more likely to be found where V(x) is small.

So the energies are different, i.e. degeneracy is broken!

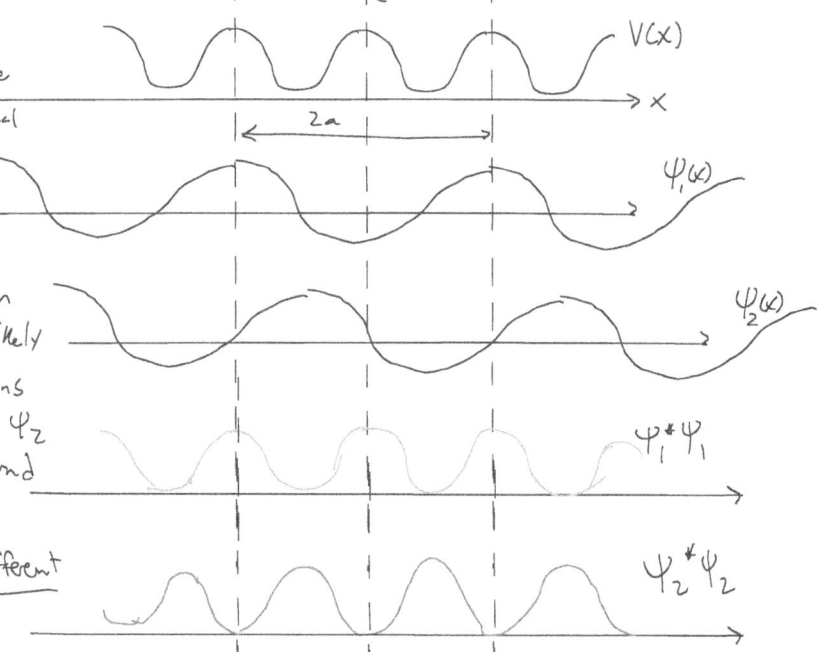

Example

$V(x) = 0.2 \cos \frac{2\pi x}{a}$ (eV), $a = 1\,nm$

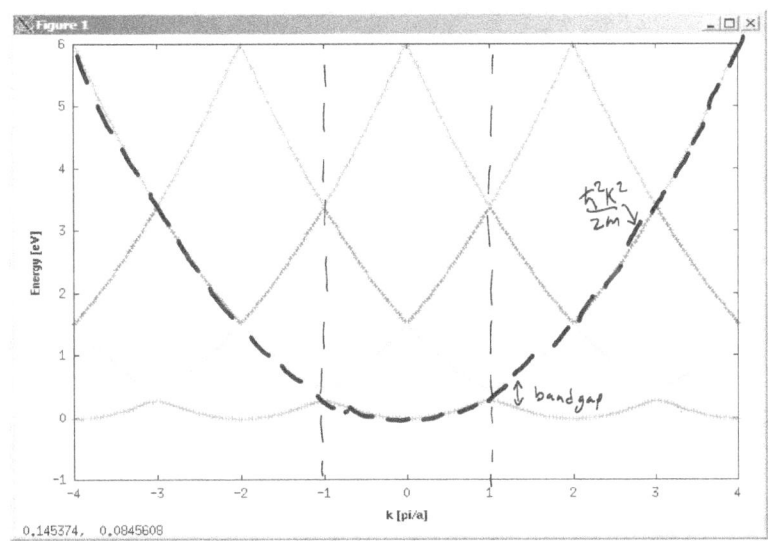

Like free electron dispersion w/ small "bandgaps" where degeneracy is lifted at Brillouin zone edge. Then, the dispersion relation is altered and so too are the dynamics!

Electrons in bands: dynamics

Consider acceleration due to a force:

$$a = \frac{dV_g}{dt} = \frac{d}{dt}\frac{d\omega}{dk} = \frac{1}{\hbar}\frac{d}{dt}\frac{dE}{dk} = \frac{1}{\hbar}\frac{d\frac{dE}{dk}}{dk}\frac{dk}{dt} = \frac{1}{\hbar^2}\frac{d^2E}{dk^2}\frac{dp}{dt}$$

$(E = \hbar\omega)$ $(p = \hbar k)$

$$= \left(\frac{1}{\hbar^2}\frac{d^2E}{dk^2}\right)F$$

$$F = \left[\frac{\hbar^2}{\frac{d^2E}{dk^2}}\right] a = m^* a$$

$m^* \to$ "effective mass", in general not equal to the rest mass, and k-dependent!

Trivial Example: plane wave $E(k) = \frac{\hbar^2(k+g)^2}{2m}$

$$m^* = \frac{\hbar^2}{\frac{d^2E}{dk^2}} = \frac{\hbar^2}{\frac{d}{dk}\frac{\hbar^2(k+g)}{m}} = m$$

Effective masses are especially affected at bandgap edges...

Bandgaps

The existence of bandgaps in the electron wave dispersion relation is critical to understanding why some crystals are metals and others are insulators or semiconductors....
Good thing we have Quantum Mechanics!

Bandstructure Examples

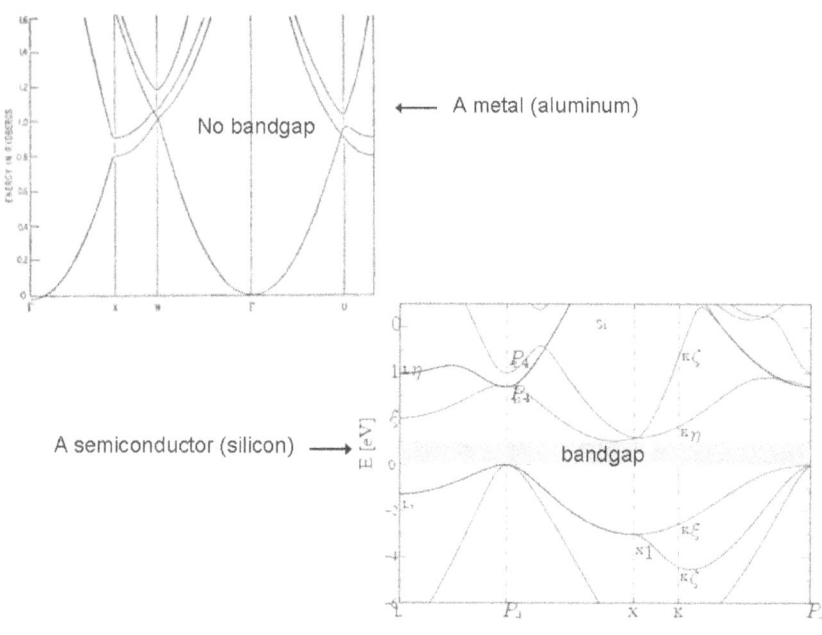

Harmonic Oscillator

$$H = \frac{p^2}{2m} + \frac{1}{2}m\omega^2 x^2 = \frac{1}{2m}\left(p^2 + (m\omega x)^2\right)$$

$V(x) = \frac{1}{2}m\omega^2 x^2$

Try to factor H:

$$H \stackrel{?}{=} \frac{1}{2m}(ip + m\omega x)(-ip + m\omega x) = \frac{1}{2m}\left(p^2 + (m\omega x)^2 + ipm\omega x + m\omega x(-ip)\right)$$

$$= \frac{1}{2m}\left(p^2 + (m\omega x)^2 + im\omega \underbrace{(px - xp)}_{[p,x]}\right)$$

Note that $[p,x]\psi = \frac{\hbar}{i}\left(\frac{d}{dx}(x\psi) - x\frac{d}{dx}\psi\right) = \frac{\hbar}{i}\left(\psi + x\frac{d\psi}{dx} - x\frac{d\psi}{dx}\right) = \frac{\hbar}{i}\psi$

So $[p,x] = \frac{\hbar}{i} = -i\hbar$ and

$$H = \frac{1}{2m}\left[(ip + m\omega x)(-ip + m\omega x) - m\hbar\omega\right]$$

$$H = \hbar\omega\left[\underbrace{\frac{1}{\sqrt{2m\hbar\omega}}(ip + m\omega x)}_{a_-}\underbrace{\frac{1}{\sqrt{2m\hbar\omega}}(-ip + m\omega x)}_{a_+} - \frac{1}{2}\right] = \hbar\omega\left(a_- a_+ - \frac{1}{2}\right)$$

40

Ladder operators

$$a_- a_+ = \frac{H}{\hbar\omega} + \frac{1}{2}$$

$$a_+ a_- = \frac{H}{\hbar\omega} - \frac{1}{2} \quad (\text{since } [p,x] = -[x,p])$$

$$[a_-, a_+] = 1$$

If ψ is an eigenfunction of H, $H\psi = E\psi$. Then, what is $a_+ \psi = ?$

$$H(a_+\psi) = \hbar\omega \left(a_+ a_- + \tfrac{1}{2}\right)(a_+\psi) = \hbar\omega \left(a_+ a_- a_+ + \tfrac{1}{2} a_+\right)\psi$$

$$= a_+ \hbar\omega \left(a_- a_+ + \tfrac{1}{2}\right)\psi = a_+ \hbar\omega \left(\frac{H}{\hbar\omega} + \tfrac{1}{2} + \tfrac{1}{2}\right)\psi$$

$$= a_+ (H + \hbar\omega)\psi = (E + \hbar\omega)(a_+\psi)$$

So $a_+\psi$ is also an eigenfunction of H, with eigenvalue greater by $\hbar\omega$! We therefore call a_+ the "raising" operator.

Ladder operators (cont)

Likewise,

$$H(a_-\psi) = \hbar\omega \left(a_- a_+ - \tfrac{1}{2}\right)(a_-\psi) = \hbar\omega \left(a_- a_+ a_- - \tfrac{1}{2} a_-\right)\psi$$

$$= a_- \hbar\omega \left(a_+ a_- - \tfrac{1}{2}\right)\psi = a_- \hbar\omega \left(\frac{H}{\hbar\omega} - \tfrac{1}{2} - \tfrac{1}{2}\right)\psi$$

$$= a_- (H - \hbar\omega)\psi = a_- (E - \hbar\omega)\psi = (E - \hbar\omega)(a_-\psi)$$

So $a_-\psi$ is also an eigenfunction of H, with eigenvalue reduced by $\hbar\omega$! We therefore call a_- the "lowering" operator.

Bottom rung of the ladder (ground state)

$$\frac{1}{\sqrt{2m\hbar\omega}}(ip + m\omega x)\psi_0 = 0$$

$$\left(\hbar\frac{d}{dx} + m\omega x\right)\psi_0 = 0$$

$$\frac{d}{dx}\psi_0 = -\frac{m\omega x}{\hbar}\psi_0$$

Normalization:
$$\int_{-\infty}^{\infty}\psi_0^* \psi_0 \, dx = \int_{-\infty}^{\infty} A^2 e^{-\frac{m\omega}{\hbar}x^2} dx$$

$$= A^2\sqrt{\frac{\pi\hbar}{m\omega}} = 1$$

$$\Rightarrow A = \left(\frac{m\omega}{\pi\hbar}\right)^{1/4}$$

This is a 1st-order differential eqn we can use to find ψ_0:

$$\int\frac{d\psi_0}{\psi_0} = -\int\frac{m\omega x}{\hbar}dx$$

$$\ln\psi_0 = -\frac{m\omega}{\hbar}\frac{x^2}{2} + C$$

$$\psi_0 = A e^{-\frac{m\omega x^2}{2\hbar}} \quad \text{gaussian "ground state"}$$

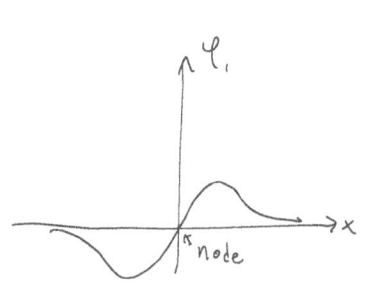
$\psi_0^*\psi_0$ lowest energy state $n=0$

Generating the first excited state

$$\psi_0 = \left(\frac{m\omega}{\pi\hbar}\right)^{1/4} e^{-\frac{m\omega x^2}{2\hbar}}$$

$$a_+ = \frac{1}{\sqrt{2m\hbar\omega}}(-ip + m\omega x)$$

$$\psi_1 = a_+\psi_0 = \frac{1}{\sqrt{2m\hbar\omega}}\left(-i\frac{\hbar}{i}\frac{d}{dx} + m\omega x\right)\left(\frac{m\omega}{\pi\hbar}\right)^{1/4} e^{-\frac{m\omega x^2}{2\hbar}}$$

$$= \left(\frac{1}{4m\omega\pi\hbar^3}\right)^{1/4}\left(\hbar\frac{m\omega}{\hbar}x + m\omega x\right) e^{-\frac{m\omega x^2}{2\hbar}}$$

$$= \left[\left(\frac{m\omega}{\hbar}\right)^3\frac{4}{\pi}\right]^{1/4} x\, e^{-\frac{m\omega x^2}{2\hbar}}$$

ψ_1, node

Note: $\int_{-\infty}^{\infty}\psi_1^*\psi_1\,dx = \frac{2}{\sqrt{\pi}}\left(\frac{m\omega}{\hbar}\right)^{3/2}\int_{-\infty}^{\infty}x^2 e^{-\frac{m\omega x^2}{\hbar}}dx = \frac{2}{\sqrt{\pi}}\xi^{3/2}\int_{-\infty}^{\infty}x^2 e^{-\xi x^2}dx = \frac{2}{\sqrt{\pi}}\xi^{3/2}\left[\frac{\sqrt{\pi}}{2\xi^{3/2}}\right] = 1$

$\left(\xi = \frac{m\omega}{\hbar}\right)$

However, higher ψ_n are not automatically normalized...

Spectrum

Since $a_- \psi_0 = 0$, $H\psi_0 = \hbar\omega\left(a_+ a_- + \frac{1}{2}\right)\psi_0 = \frac{\hbar\omega}{2}\psi_0$.

Higher eigenvalues are spaced by $\hbar\omega$, so we have

$$E_n = \left(n + \frac{1}{2}\right)\hbar\omega, \quad n = 0, 1, 2, \ldots$$

(This is equivalent to recognizing that the eigenvalues of $a_+ a_-$ are $n = 0, 1, 2 \ldots$)

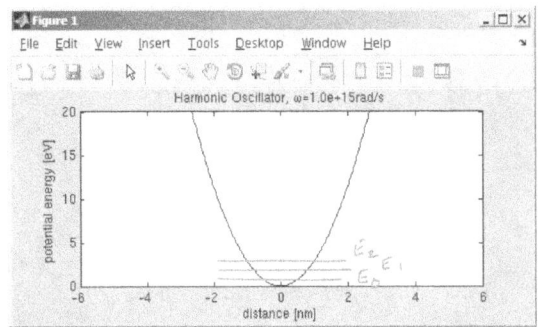

Numerical Solution

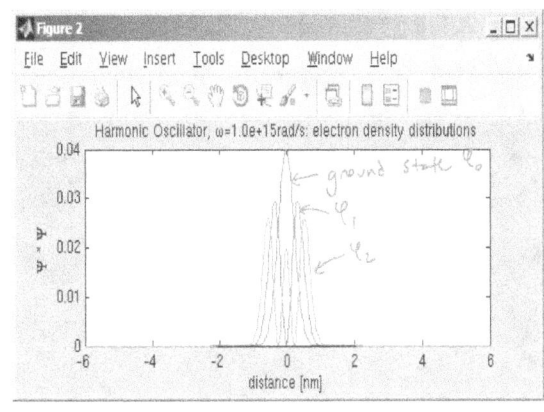

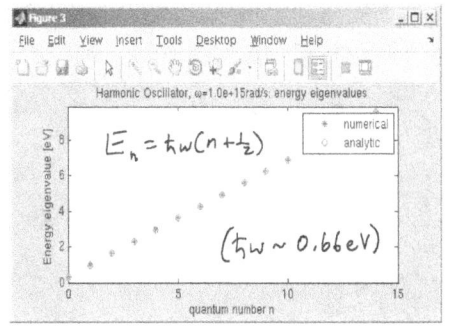

Notice that the probability density $\psi^*\psi$ is confined to the bottom of the well, decays into classically forbidden region, and acquires more nodes (where $\psi^*\psi = 0$) as n increases.

Ground-state properties from the Heisenberg uncertainty relation

Energy of a wavefunction is determined by the expectation value of Hamiltonian for Harmonic Oscillator $\langle H \rangle = \frac{\langle p^2 \rangle}{2m} + \frac{1}{2}m\omega^2 \langle x^2 \rangle$.

Since $\langle p \rangle = 0$ for bound state and $\langle x \rangle = 0$ for symmetric potential,

$$\Delta x = \sqrt{\langle x^2 \rangle - \cancelto{0}{\langle x \rangle^2}} \rightarrow \langle x^2 \rangle = \Delta x^2 \quad \text{and} \quad \Delta p = \sqrt{\langle p^2 \rangle - \cancelto{0}{\langle p \rangle^2}} \rightarrow \langle p^2 \rangle = \Delta p^2$$

So $\langle H \rangle = \frac{\Delta p^2}{2m} + \frac{1}{2}m\omega^2 \Delta x^2$ but we don't know either Δx or Δp.

However, we know the two are connected by the Heisenberg rel'n:

$$\Delta x \Delta p \geq \frac{\hbar}{2}. \quad \text{Using } \Delta p \geq \frac{\hbar}{2\Delta x}, \text{ we have}$$

$$\langle H \rangle \leq \frac{\hbar^2}{8m\Delta x^2} + \frac{1}{2}m\omega^2 \Delta x^2$$

The ground-state energy must be lower than the minimum value of the RHS.

Minimization

$$\frac{d}{d\Delta x}\left(\frac{\hbar^2}{8m\Delta x^2} + \frac{1}{2}m\omega^2 \Delta x^2\right) = -\frac{\hbar^2}{4m\Delta x^3} + m\omega^2 \Delta x = 0 \quad \text{which has sol'n } \Delta x = \sqrt{\frac{\hbar}{2m\omega}}$$

Let's compare to the exact solution $\psi_0 \propto e^{-\frac{m\omega x^2}{2\hbar}}$!

probability density $\psi_0^* \psi_0 \propto e^{-\frac{m\omega x^2}{\hbar}} = e^{-\frac{x^2}{2\Delta x^2}}$ where $\Delta x = \sqrt{\frac{\hbar}{2m\omega}}$ — Same!

We can also calculate the upper bound on energy of the ground state:

$$\langle H \rangle \leq \frac{\hbar^2}{8m\Delta x^2} + \frac{1}{2}m\omega^2 \Delta x^2 = \frac{\hbar^2}{8m \frac{\hbar}{2m\omega}} + \frac{1}{2}m\omega^2 \frac{\hbar}{2m\omega} = \frac{\hbar\omega}{4} + \frac{\hbar\omega}{4} = \frac{\hbar\omega}{2} \quad \text{Same as analytic value for } E_0!$$

In this case, the inequality is an **equality** because the gaussian ground state minimizes the product $\Delta x \Delta p$!

In general, of course, ground state wavefunctions are not gaussian and this procedure only gives an upper bound.

Classical harmonic oscillator

$F = -kx \longrightarrow V(x) = -\int F(x)dx = \frac{1}{2}kx^2$

Newton's 2nd Law: $F = ma = m\ddot{x}$

$\ddot{x}(t) = -\frac{k}{m}x(t)$ Sol'n: $x(t) = A\sin\omega t + B\cos\omega t$, $\omega = \sqrt{\frac{k}{m}} \to k = m\omega^2$

Let's arbitrarily choose initial conditions s.t. $B=0$. Then $A = x_{max}$,
the "classical turning point" where $E = V(x) = \frac{m\omega^2}{2}x_{max}^2 \longrightarrow x_{max} = \sqrt{\frac{2E}{m\omega^2}}$

Classical probability density can be obtained from this eqn. of motion by calculating $dt(x)$:

$x(t) = x_{max} \sin \omega t$

$t(x) = \dfrac{a\sin \frac{x}{x_{max}}}{\omega}$

Intuitively, $dt(x)$ larger near x_{max}, because mass is moving slower there. Graphically, slope is lower.

$dt < dt$

Classical probability distribution

$dt(x) = \dfrac{dt}{dx}dx = \dfrac{d}{dx}\left(\dfrac{a\sin\frac{x}{x_{max}}}{\omega}\right) \cdot dx = \dfrac{dx}{\omega x_{max}\sqrt{1-\left(\frac{x}{x_{max}}\right)^2}} = \dfrac{dx}{\omega\sqrt{x_{max}^2 - x^2}}$

OR

$= \dfrac{1}{\frac{dx}{dt}}dx = \dfrac{dx}{x_{max}\omega\cos\omega t} = \dfrac{dx}{x_{max}\omega\cos(a\sin\frac{x}{x_{max}})}$

$= \dfrac{dx}{x_{max}\omega \cdot \frac{\sqrt{x_{max}^2-x^2}}{x_{max}}} = \dfrac{dx}{\omega\sqrt{x_{max}^2-x^2}}$ Same!

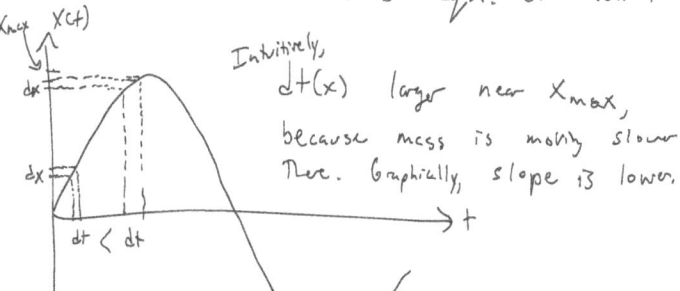

Probability density is $dt(x)$ normalized by half a period per dx

$$P_{classical} = \dfrac{1}{\pi\sqrt{x_{max}^2 - x^2}} \quad \Longleftrightarrow \quad P_{quantum} = \psi^*\psi$$

Numerical comparison

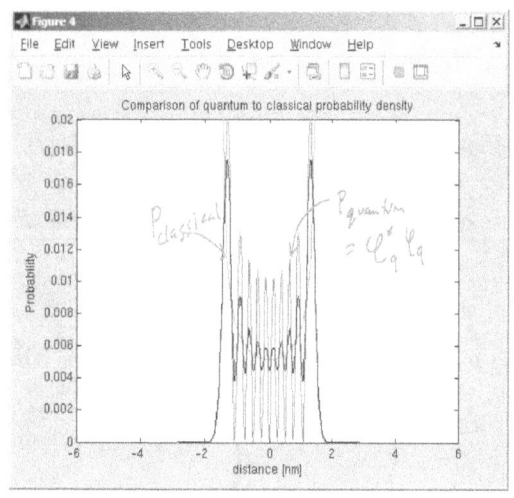 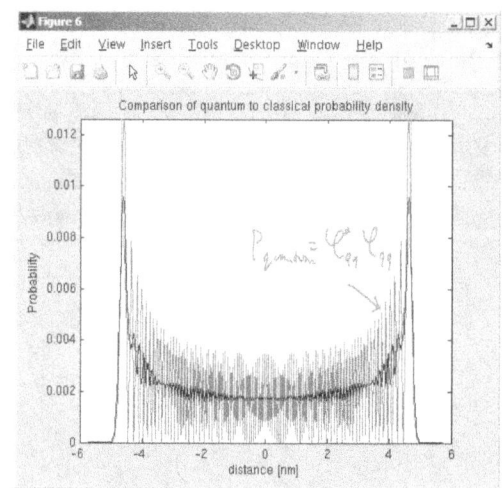

We can see the "correspondence principle" working here: Quantum systems behave like their classical counterparts in the limit $n \to \infty$!

A more direct approach

$$-\frac{\hbar^2}{2m}\frac{d^2\varphi}{dx^2} + \frac{1}{2}m\omega^2 x^2 \varphi = E\varphi$$

divide by $-\frac{\hbar\omega}{2}$

$$\frac{\hbar}{m\omega}\frac{d^2\varphi}{dx^2} - \frac{m\omega}{\hbar}x^2\varphi = -\frac{2E}{\hbar\omega}\varphi$$

define $\zeta = \sqrt{\frac{m\omega}{\hbar}}x$ (unitless) such that $x = \sqrt{\frac{\hbar}{m\omega}}\zeta$ and $dx = \sqrt{\frac{\hbar}{m\omega}}d\zeta$:

$$\frac{d^2\varphi}{d\zeta^2} = (\zeta^2 - K)\varphi$$

$\left(K \equiv \frac{2E}{\hbar\omega}\right)$. So far, we have only recast our Schrödinger eqn in a more manageable, unitless form. We haven't gotten any closer to solving it tho!

Asymptotic behavior

For large $\xi \gg 1$ (large x)

$$\frac{d^2\psi}{d\xi^2} \approx \xi^2 \psi$$

This has approximate sol'n $\psi(\xi) = Ae^{-\xi^2/2} + Be^{+\xi^2/2}$ $=0$ so that ψ is normalizable.

Check it: $\frac{d\psi}{d\xi} = -\xi A e^{-\xi^2/2}$, $\frac{d^2\psi}{d\xi^2} = A\left(\xi^2 e^{-\xi^2/2} - e^{-\xi^2/2}\right)$

$$A\left(\xi^2 e^{-\xi^2/2} - \underbrace{e^{-\xi^2/2}}_{\text{negligible for } \xi \gg 1}\right) \approx \xi^2 A e^{-\xi^2/2}$$

This suggests that all our solutions will be proportional to $e^{-\xi^2/2}$, since it will dominate for large ξ.

Ansatz $\psi(\xi) = h(\xi) e^{-\xi^2/2}$

$$\frac{d\psi}{d\xi} = h' e^{-\xi^2/2} - \xi e^{-\xi^2/2} h$$

$$\frac{d^2\psi}{d\xi^2} = h'' e^{-\xi^2/2} - \xi e^{-\xi^2/2} h' - \left(h' \xi e^{-\xi^2/2} + h\left(e^{-\xi^2/2} - \xi^2 e^{-\xi^2/2}\right)\right)$$

Schrödinger eqn becomes:

$$h'' e^{-\xi^2/2} - \xi e^{-\xi^2/2} h' - \left(h' \xi e^{-\xi^2/2} + h\left(e^{-\xi^2/2} - \xi^2 e^{-\xi^2/2}\right)\right) = (\xi^2 - K) h e^{-\xi^2/2}$$

After eliminating common factor of $e^{-\xi^2/2}$,

$$\frac{d^2 h}{d\xi^2} - 2\xi \frac{dh}{d\xi} + (K-1) h = 0$$

Brute force

Substitute: $h(\xi) = \sum_{j=0}^{\infty} a_j \xi^j$ so $\frac{dh}{d\xi} = \sum_{j=0}^{\infty} j a_j \xi^{j-1}$

and $\frac{d^2h}{d\xi^2} = \sum_{j=0}^{\infty} j(j-1) a_j \xi^{j-2} = \sum_{j=2}^{\infty} j(j-1) a_j \xi^{j-2} = \sum_{j=0}^{\infty} (j+2)(j+1) a_{j+2} \xi^j$

Then our eqn for $h(\xi)$ becomes:

$$\sum_{j=0}^{\infty} (j+2)(j+1) a_{j+2} \xi^j - \sum_{j=0}^{\infty} 2\xi j a_j \xi^{j-1} + \sum_{j=0}^{\infty} (K-1) a_j \xi^j = 0$$

$$\sum_{j=0}^{\infty} \left[(j+2)(j+1) a_{j+2} - 2j a_j + (K-1) a_j \right] \xi^j = 0$$

This can only be true if <u>all</u> coefficients of all terms are zero!

Recursion relation

The coefficients can be written $(j+2)(j+1) a_{j+2} - (2j + 1 - K) a_j = 0$

This leads to: $a_{j+2} = \frac{2j+1-K}{(j+2)(j+1)} a_j$ "recursion relation"

For large $j \gg 1$, $a_{j+2} \approx \frac{2}{j} a_j \Rightarrow a_j \approx \frac{C}{(j/2)!}$

This gives $h(\xi) \approx \sum_{j=0}^{\infty} \frac{C}{(j/2)!} \xi^j \approx \sum_{j=0}^{\infty} C \frac{(\xi^2)^j}{j!} = C e^{\xi^2}$ But, this is not normalizable!

So series must terminate for some $j_{max} = n$ so that $a_{j_{max}+2} = 0$

i.e. $h(\xi)$ is a polynomial ("Hermite polynomial")

$$2n+1-K=0 \Rightarrow 2n+1 - \frac{2E}{\hbar\omega} = 0$$

$$E = \hbar\omega\left(n+\tfrac{1}{2}\right), \quad n = 0, 1, 2, \ldots$$

Same as ladder operator method! ✓

Wavefunctions

For an eigenfunction, $K = 2n+1$. Therefore,

$$a_{j+2} = \frac{2j+1 - (2n+1)}{(j+1)(j+2)} a_j = \frac{-2(n-j)}{(j+1)(j+2)} a_j, \quad j \leq n$$

We want to generate $h_n(\zeta) = \sum_{j}^{\infty} a_j \zeta^j$, and wavefunctions $\psi_n(\zeta) = h_n(\zeta) e^{-\zeta^2/2}$

for $n=0$, $h_0(\zeta) = a_0$, $\psi_0(\zeta) \propto e^{-\zeta^2/2}$

$n=1$, $h_1(\zeta) = a_1 \zeta$, $\psi_1(\zeta) \propto \zeta e^{-\zeta^2/2}$

$n=2$, for $j=0$, $a_2 = -2a_0$ so $h_2(\zeta) = a_0 - 2a_0 \zeta^2$, $\psi_2(\zeta) \propto (1 - 2\zeta^2) e^{-\zeta^2/2}$

etc....

Scattering States

$E > V(x)$ as $x \to \pm\infty$

$\psi_- = \psi_{inc} + \psi_{refl.} = e^{iK_1 x} + r e^{-iK_1 x}$, $K_1 = \sqrt{\frac{2mE}{\hbar^2}}$, $V(x<0) = 0$

$\psi_+ = \psi_{trans} = t e^{iK_2 x}$, $K_2 = \sqrt{\frac{2m(E-V_0)}{\hbar^2}}$, $V(x>0) = V_0$

How much probability flux is transmitted / reflected?

Piecewise-constant $V(x)$: Solutions to time-independent Schrödinger eqn are planewaves.

$$-\frac{\hbar^2}{2m} \psi'' + V(x)\psi = E\psi \Rightarrow \psi'' = -k^2 \psi, \quad \left(k = \sqrt{\frac{2m(E-V)}{\hbar^2}}\right)$$

Boundary Conditions

① $\psi_-(0) = \psi_+(0) \to 1 + r = t$

② $\psi_-'(0) = \psi_+'(0) \to iK_1 - iK_1 r = iK_2 t = iK_2(1+r) \to K_1 - K_1 r = K_2 + K_2 r$

$\to r = \frac{K_1 - K_2}{K_1 + K_2}$ Then $t = 1 + r = \frac{K_1 + K_2}{K_1 + K_2} + \frac{K_1 - K_2}{K_1 + K_2} = \frac{2K_1}{K_1 + K_2}$

Transmission and Reflection coefficients

$$J = \frac{\hbar}{m} \text{Im}\left\{\psi^* \frac{d}{dx}\psi\right\}$$

For planewave αe^{ikx}, $J = \frac{\hbar k}{m} \psi^* \psi$

transmission: $T = \dfrac{J_{trans}}{J_{inc}} = \dfrac{\frac{\hbar k_2}{m}(t^* e^{-ik_2 x} \cdot t e^{ik_2 x})}{\frac{\hbar k_1}{m}(e^{-ik_1 x} \cdot e^{ik_1 x})} = \dfrac{k_2}{k_1}|t|^2 = \dfrac{k_2}{k_1} \dfrac{4k_1^2}{(k_1+k_2)^2} = \dfrac{4k_1 k_2}{(k_1+k_2)^2}$

reflection: $R = \dfrac{J_{refl}}{J_{inc}} = \dfrac{\frac{\hbar k_1}{m}|r|^2}{\frac{\hbar k_1}{m}} = |r|^2 = \dfrac{(k_1-k_2)^2}{(k_1+k_2)^2}$

$T + R = \dfrac{4k_1 k_2 + k_1^2 - 2k_1 k_2 + k_2^2}{(k_1+k_2)^2} = \dfrac{(k_1+k_2)^2}{(k_1+k_2)^2} = 1 \checkmark$

Transmission vs. Energy

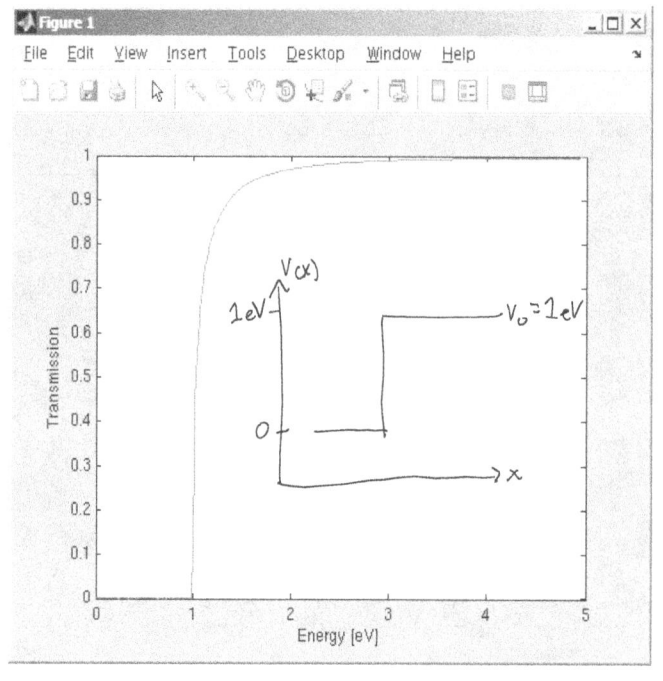

Note $T < 1$ above barrier!
(classical mechanics would of course give $T = 1$ for $E > V_0$)

Scattering from divergent potentials $V(x) = -\alpha\delta(x)$

I. Determine $\psi(x) = \begin{cases} \psi_-(x) = e^{ikx} + re^{-ikx}, & x<0 \\ \psi_+(x) = t e^{ikx}, & x>0 \end{cases}$

$\left(k = \sqrt{\frac{2mE}{\hbar^2}}\right)$

II. Apply B.C.'s to each interface

① $\psi_-(0) = \psi_+(0)$

② $\psi'_+(0) - \psi'_-(0) = -\frac{2m\alpha}{\hbar^2}\psi(0)$ (from integration of T.I.S.E.)

III. Calculate $T = J_{trans}/J_{inc}$

Applying boundary conditions

① $1 + r = t$

② $ikt - (ik - ikr) = -\frac{2m\alpha}{\hbar^2}(1+r)$

$ik(1+r) - ik(1-r) = -\frac{2m\alpha}{\hbar^2}(1+r)$

$2ikr = -\frac{2m\alpha}{\hbar^2}(1+r)$

$r\left(ik + \frac{m\alpha}{\hbar^2}\right) = -\frac{m\alpha}{\hbar^2}$

$r = -\frac{\frac{m\alpha}{\hbar^2}}{ik + \frac{m\alpha}{\hbar^2}} = \frac{\frac{m\alpha}{\hbar^2 k}i}{1 - \frac{m\alpha}{\hbar^2 k}i} = \frac{i\beta}{1-i\beta}$ $\left(\beta = \frac{m\alpha}{\hbar^2 k}\right)$

$t = 1 + r = \frac{(1-i\beta) + i\beta}{1-i\beta} = \frac{1}{1-i\beta}$

III. $T = \frac{\frac{\hbar k}{m}|t|^2}{\frac{\hbar k}{m}} = |t|^2 = \frac{1}{1+i\beta}\frac{1}{1-i\beta} = \frac{1}{1+\beta^2}$

Due to dependence on α^2, same $T(E)$ for repulsive δ-fn!

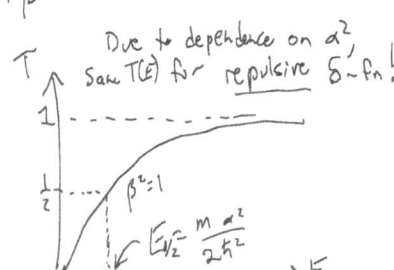

>1 interface

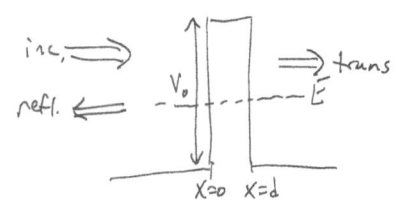

for $x<0$ and $x>d$:
$$-\frac{\hbar^2}{2m}\psi'' = E\psi \implies \psi'' = -k^2\psi, \quad k = \sqrt{\frac{2mE}{\hbar^2}}$$

for $0<x<d$:
$$-\frac{\hbar^2}{2m}\psi'' + V_0\psi = E\psi \implies \psi'' = \kappa^2\psi, \quad \kappa = \sqrt{\frac{2m(V_0-E)}{\hbar^2}}$$

I. find $\psi(x)$ in all three regions
$$\psi(x) = \begin{cases} \psi_-(x) = e^{ikx} + re^{-ikx}, & x<0 \\ \psi_b(x) = Ae^{\kappa x} + Be^{-\kappa x}, & 0<x<d \\ \psi_+(x) = te^{ikx}, & x>d \end{cases}$$

II. Apply Boundary conditions

① $\psi_-(0) = \psi_b(0)$ ③ $\psi_b(d) = \psi_+(d)$

② $\psi_-'(0) = \psi_b'(0)$ ④ $\psi_b'(d) = \psi_+'(d)$

{ But this is tedious and not practical for complicated potentials w/ many interfaces! }

III. Calculate $T = J_{trans}/J_{inc}$

Boundary conditions at an arbitrary interface

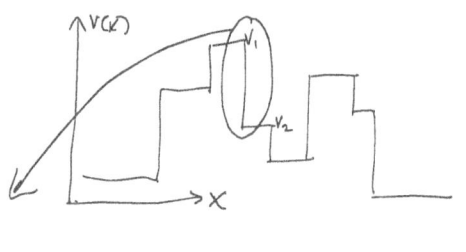

$Ae^{ik_1x} \Rightarrow$ | $Ce^{ik_2x} \Rightarrow$

$+Be^{-ik_1x} \Leftarrow$ | $+De^{-ik_2x} \Leftarrow$

$k_1 = \sqrt{\frac{2m(E-V_1)}{\hbar^2}}$ | $k_2 = \sqrt{\frac{2m(E-V_2)}{\hbar^2}}$

bdry

ψ continuous: $Ae^{ik_1x} + Be^{-ik_1x}\big|_{bdry} = Ce^{ik_2x} + De^{-ik_2x}\big|_{bdry}$

ψ' continuous: $ik_1 Ae^{ik_1x} - ik_1 Be^{-ik_1x}\big|_{bdry} = ik_2 Ce^{ik_2x} - ik_2 De^{-ik_2x}\big|_{bdry}$

A matrix equation:
$$\begin{bmatrix} e^{ik_1x} & e^{-ik_1x} \\ ik_1 e^{ik_1x} & -ik_1 e^{-ik_1x} \end{bmatrix} \begin{bmatrix} A \\ B \end{bmatrix}\bigg|_{bdry} = \begin{bmatrix} e^{ik_2x} & e^{-ik_2x} \\ ik_2 e^{ik_2x} & -ik_2 e^{-ik_2x} \end{bmatrix} \begin{bmatrix} C \\ D \end{bmatrix}\bigg|_{bdry}$$

Boundary conditions at an arbitrary interface (cont)

$$\begin{bmatrix} A \\ B \end{bmatrix} = \begin{bmatrix} \frac{1}{2} e^{-ik_1 x} & -\frac{1}{2}\frac{i}{k_1} e^{-ik_1 x} \\ \frac{1}{2} e^{ik_1 x} & \frac{1}{2}\frac{i}{k_1} e^{ik_1 x} \end{bmatrix} \begin{bmatrix} e^{ik_2 x} & e^{-ik_2 x} \\ ik_2 e^{ik_2 x} & -ik_2 e^{-ik_2 x} \end{bmatrix}_{bdry} \begin{bmatrix} C \\ D \end{bmatrix}$$

$$\begin{bmatrix} A \\ B \end{bmatrix} = \begin{bmatrix} \left(\frac{1}{2} + \frac{k_2}{2k_1}\right) e^{i(k_2 - k_1)x} & \left(\frac{1}{2} - \frac{k_2}{2k_1}\right) e^{-i(k_1 + k_2)x} \\ \left(\frac{1}{2} - \frac{k_2}{2k_1}\right) e^{i(k_1 + k_2)x} & \left(\frac{1}{2} + \frac{k_2}{2k_1}\right) e^{-i(k_2 - k_1)x} \end{bmatrix}_{x = bdry.} \begin{bmatrix} C \\ D \end{bmatrix}$$

For arbitrary potential:

$$k_L \begin{bmatrix} A \\ B \end{bmatrix} \underbrace{\begin{bmatrix} C \\ D \end{bmatrix}}_{1} \underbrace{\begin{bmatrix} E \\ F \end{bmatrix}}_{2} \underbrace{\begin{bmatrix} G \\ H \end{bmatrix}}_{3} \cdots \underbrace{\begin{bmatrix} U \\ V \end{bmatrix}}_{N-2} \underbrace{\begin{bmatrix} W \\ X \end{bmatrix}}_{N-1} \underbrace{\begin{bmatrix} Y \\ Z \end{bmatrix}}_{N} k_R$$

$$\begin{bmatrix} A \\ B \end{bmatrix} = \begin{bmatrix} \text{interface 1} \end{bmatrix} \cdots \begin{bmatrix} \text{interface N-1} \end{bmatrix} \begin{bmatrix} \text{interface N} \end{bmatrix} \begin{bmatrix} Y \\ Z \end{bmatrix} = \overleftrightarrow{M} \begin{bmatrix} Y \\ Z \end{bmatrix}$$

Transmission coefficient

$$\begin{bmatrix} A = 1 \\ B = r \end{bmatrix} = \begin{bmatrix} M_{11} & M_{12} \\ M_{21} & M_{22} \end{bmatrix} \begin{bmatrix} Y = t \\ Z = 0 \end{bmatrix}$$

$$1 = M_{11} t \implies t = \frac{1}{M_{11}}$$

$$r = M_{21} t \implies r = \frac{M_{21}}{M_{11}}$$

$$T = \frac{J_{trans}}{J_{inc}} = \frac{\frac{\hbar k_R}{m} |t|^2}{\frac{\hbar k_L}{m}} = \frac{k_R}{k_L} \left|\frac{1}{M_{11}}\right|^2 , \qquad R = \left|\frac{M_{21}}{M_{11}}\right|^2$$

A recipe

For a given piecewise-constant scattering potential:

1. Pick E
2. Construct 2×2 matrix for each interface from $E, V_1, V_2,$ and position of interface x
3. Multiply them together (in proper order) to get $\hat{M}$
4. Calculate $T(E) = \frac{K_R}{K_L} \frac{1}{|M_{11}|^2}$
5. Go to #1 (repeat for different E)

Example: step potential

$K_1 = \sqrt{\frac{2mE}{\hbar^2}}$, $K_2 = \sqrt{\frac{2m(E-V_0)}{\hbar^2}}$, at $x=0$

$$M = \begin{pmatrix} \left(\frac{1}{2}+\frac{K_2}{2K_1}\right)e^{i(K_2-K_1)x} & \left(\frac{1}{2}-\frac{K_2}{2K_1}\right)e^{-i(K_2+K_1)x} \\ \left(\frac{1}{2}-\frac{K_2}{2K_1}\right)e^{i(K_2+K_1)x} & \left(\frac{1}{2}+\frac{K_2}{2K_1}\right)e^{i(K_1-K_2)x} \end{pmatrix} = \begin{pmatrix} \frac{1}{2}+\frac{K_2}{2K_1} & \frac{1}{2}-\frac{K_2}{2K_1} \\ \frac{1}{2}-\frac{K_2}{2K_1} & \frac{1}{2}+\frac{K_2}{2K_1} \end{pmatrix}$$

So $M_{11} = \frac{1}{2}+\frac{K_2}{2K_1}$

$$T = \frac{K_R}{K_L}\frac{1}{|M_{11}|^2} = \frac{K_2}{K_1}\frac{1}{\left(\frac{1}{2}+\frac{K_2}{2K_1}\right)^2} = \frac{4K_2}{K_1\left(1+\frac{K_2}{K_1}\right)^2} = \frac{4K_1 K_2}{K_1^2\left(1+\frac{K_2}{K_1}\right)^2} = \frac{4K_1 K_2}{(K_1+K_2)^2}$$

c.f. previous result

single barrier

[Diagram: rectangular barrier of height V_0 with regions I, II, III; II spans $x=0$ to $x=d$]

What is $T(E)$?

$$\left(K_I = K_{III} = \sqrt{\frac{2mE}{\hbar^2}}, \quad K_{II} = \sqrt{\frac{2m(E-V_0)}{\hbar^2}} \right)$$

for each interface, construct

$$\begin{bmatrix} \left(\frac{1}{2}+\frac{k_2}{2k_1}\right)e^{i(k_2-k_1)x} & \left(\frac{1}{2}-\frac{k_2}{2k_1}\right)e^{-i(k_2+k_1)x} \\ \left(\frac{1}{2}-\frac{k_2}{2k_1}\right)e^{i(k_2+k_1)x} & \left(\frac{1}{2}+\frac{k_2}{2k_1}\right)e^{i(k_1-k_2)x} \end{bmatrix}$$

(where k_1 is to left of interface, k_2 is to right.)

$$\begin{bmatrix} \left(\frac{1}{2}+\frac{K_{II}}{2K_I}\right) & \left(\frac{1}{2}-\frac{K_{II}}{2K_I}\right) \\ \Box & \Box \end{bmatrix} \begin{bmatrix} \left(\frac{1}{2}+\frac{K_{III}}{2K_{II}}\right)e^{i(K_{III}-K_{II})d} & \Box \\ \left(\frac{1}{2}-\frac{K_{III}}{2K_{II}}\right)e^{i(K_{III}+K_{II})d} & \Box \end{bmatrix} = \begin{bmatrix} M_{11} & M_{12} \\ M_{21} & M_{22} \end{bmatrix}$$

@ $x=0$ — only need these elements! — @ $x=d$

single barrier (cont)

$$M_{11} = \left(\frac{1}{2}+\frac{K_{II}}{2K_I}\right)\left(\frac{1}{2}+\frac{K_{III}}{2K_{II}}\right)e^{i(K_{III}-K_{II})d} + \left(\frac{1}{2}-\frac{K_{II}}{2K_I}\right)\left(\frac{1}{2}-\frac{K_{III}}{2K_{II}}\right)e^{i(K_{III}+K_{II})d}$$

$$= e^{i(K_{III}-K_{II})d}\left(A + B e^{+2iK_{II}d}\right)$$

$$T = \left|\frac{1}{M_{11}}\right|^2 = \frac{1}{A^2+B^2+AB(e^{2iK_{II}d}+e^{-2iK_{II}d})} = \frac{1}{A^2+B^2+2AB\cos 2K_{II}d}$$

↖ oscillating if K_{II} real ($E > V_0$)

Note asymptotic behavior:

If $E \gg V_0$, $K_I, K_{III} \sim K_{II}$ so $A \to 1$ and $B \to 0$ so $T \to 1$

If E small, $K_I, K_{III} \to 0$ so A, B diverge and $T \to 0$

Results

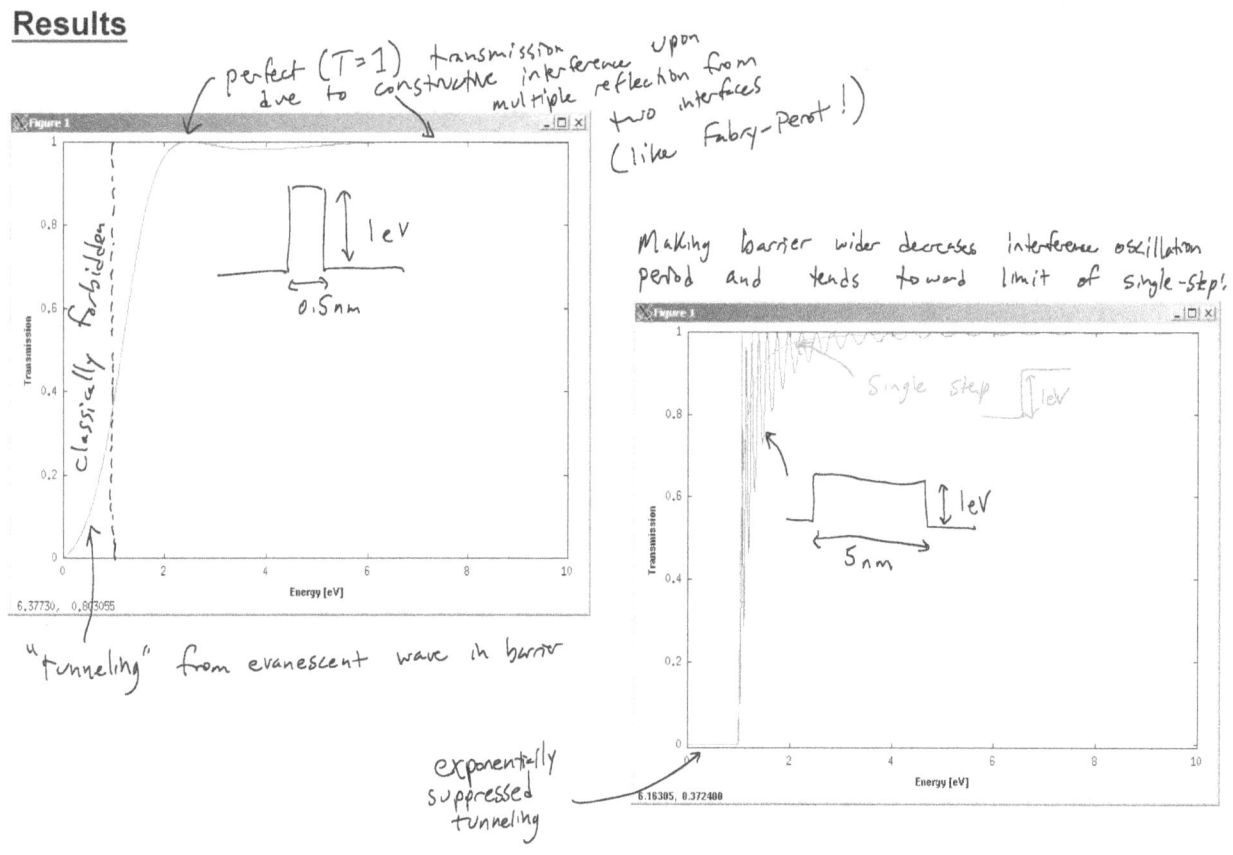

- perfect (T=1) transmission due to constructive interference upon multiple reflection from two interfaces (like Fabry-Perot!)
- classically forbidden
- "tunneling" from evanescent wave in barrier
- exponentially suppressed tunneling
- Making barrier wider decreases interference oscillation period and tends toward limit of single-step!
- Single step

Two barriers

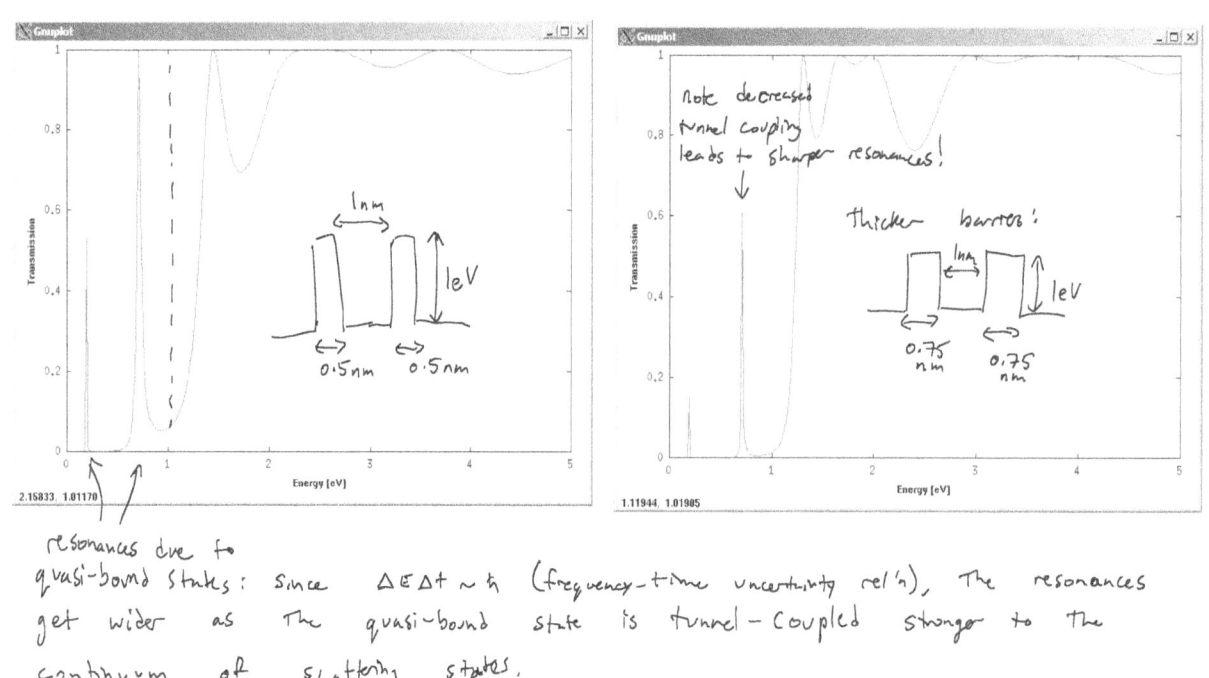

- resonances due to quasi-bound states: since $\Delta E \Delta t \sim \hbar$ (frequency-time uncertainty rel'n), the resonances get wider as the quasi-bound state is tunnel-coupled stronger to the continuum of scattering states.
- note decreased tunnel coupling leads to sharper resonances!
- thicker barriers

Three barriers

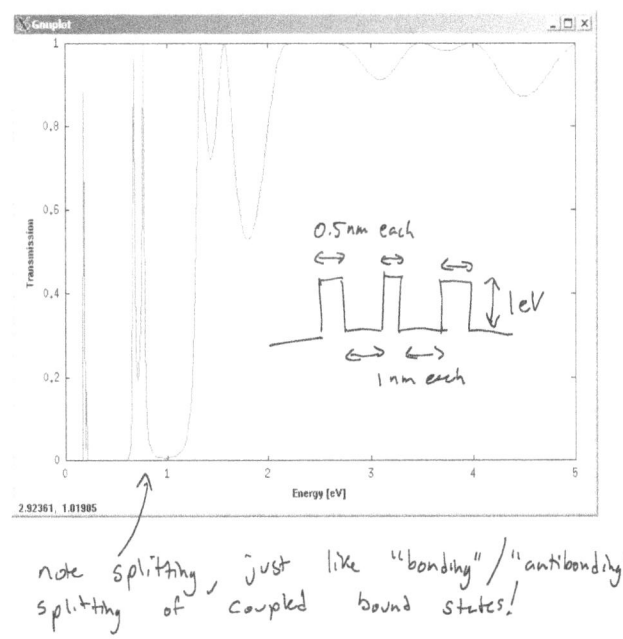

note splitting, just like "bonding"/"antibonding" splitting of coupled bound states!

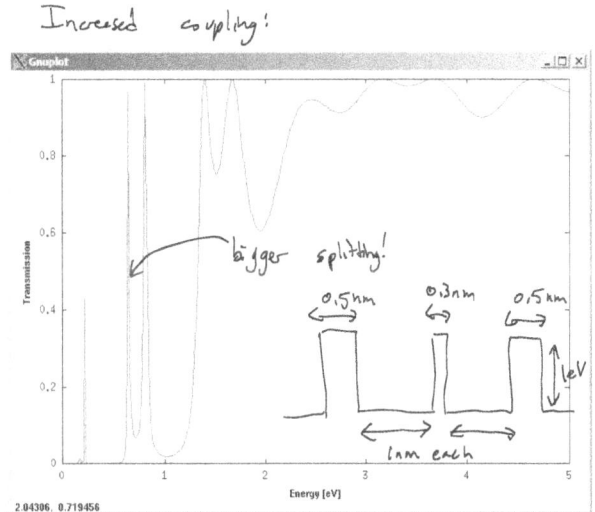

Increased coupling!

bigger splitting!

Four barriers and beyond

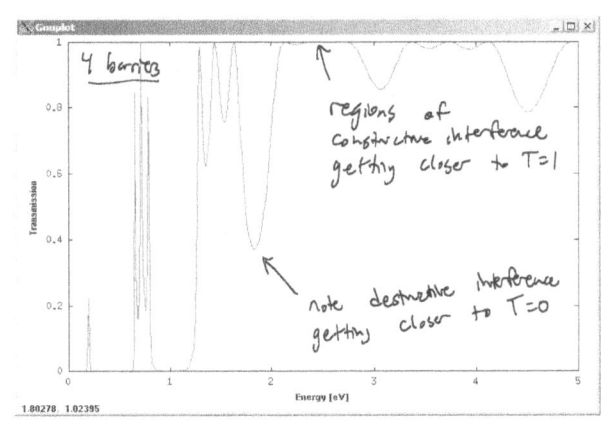

4 barriers

regions of constructive interference getting closer to T=1

note destructive interference getting closer to T=0

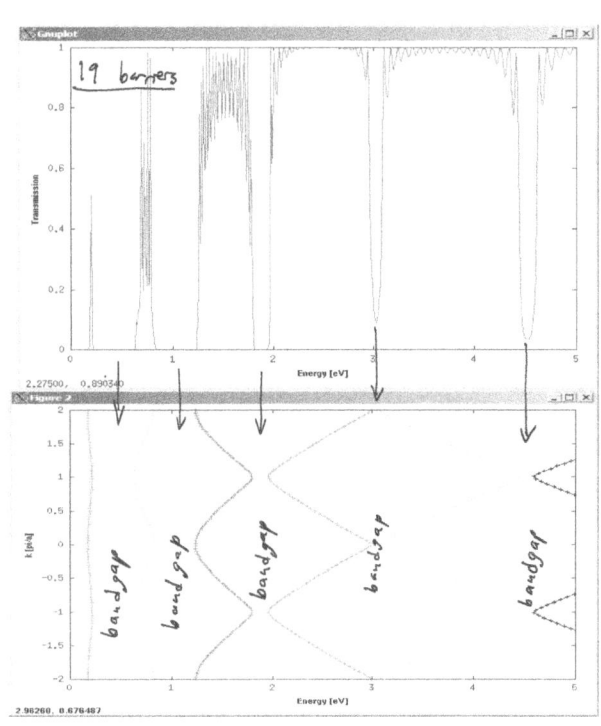

19 barriers

bandgap

In limit of infinite barriers, we have full translational symmetry and k is conserved → "bands" and "bandstructure"!

Example: parabolic scatterer

$$V(x) = \begin{cases} 0, & x < -4nm \\ \alpha x^2, & -4nm < x < 4nm \\ 0, & x > 4nm \end{cases}$$

$\alpha = 10^{14} \, eV/cm^2$

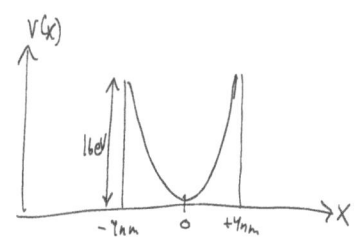

Qualitatively predict $T(E)$ for $E < 16 eV$ (below top of barriers):

Tunneling into quasi-bound states similar to $V(x) = \frac{1}{2} m \omega^2 x^2$ (SHO).

Bound states of SHO are $E = \hbar \omega (n + \frac{1}{2})$ $n = 0, 1, 2, \ldots$

Where $\alpha = \frac{1}{2} m \omega^2 \rightarrow \omega = \sqrt{\frac{2\alpha}{m}} \sim \sqrt{\frac{2 \times 10^{14} \, eV/cm^2}{5.7 \times 10^{-16} \, eV s^2/cm^2}} \sim 5.9 \times 10^{14} \, rad/s \rightarrow \hbar \omega = 0.4 eV$

So we expect a series of resonances equally spaced by $\sim 0.4 eV$. Each resonance will get wider as E increases due to increased coupling to propagating states.

Numerical result

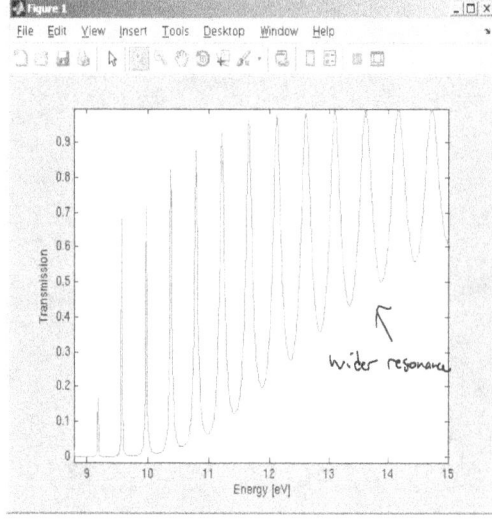
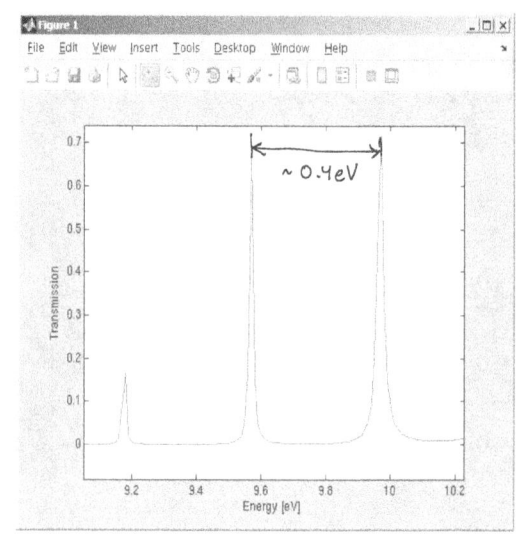

Scattering: finite differences

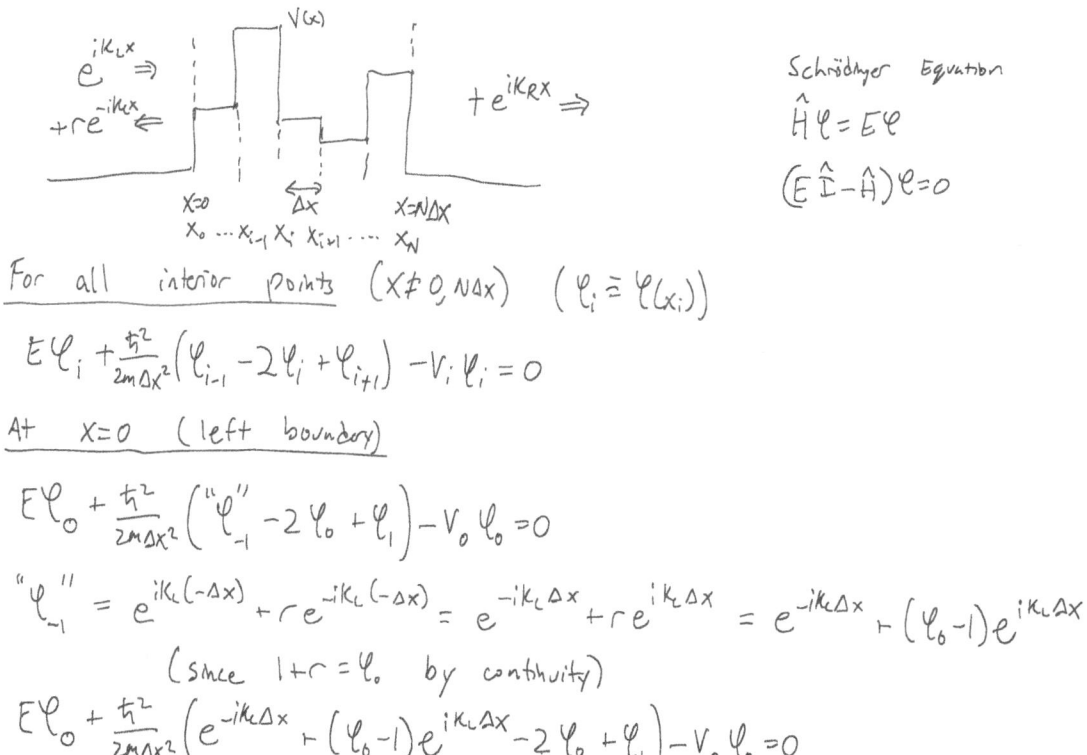

Schrödinger Equation

$$\hat{H}\psi = E\psi$$
$$(E\hat{I} - \hat{H})\psi = 0$$

For all interior points $(x \neq 0, N\Delta x)$ $(\psi_i \equiv \psi(x_i))$

$$E\psi_i + \frac{\hbar^2}{2m\Delta x^2}(\psi_{i-1} - 2\psi_i + \psi_{i+1}) - V_i \psi_i = 0$$

At $x=0$ (left boundary)

$$E\psi_0 + \frac{\hbar^2}{2m\Delta x^2}("\psi"_{-1} - 2\psi_0 + \psi_1) - V_0 \psi_0 = 0$$

$$"\psi"_{-1} = e^{ik_L(-\Delta x)} + re^{-ik_L(-\Delta x)} = e^{-ik_L\Delta x} + re^{ik_L\Delta x} = e^{-ik_L\Delta x} + (\psi_0 - 1)e^{ik_L\Delta x}$$

(since $1 + r = \psi_0$ by continuity)

$$E\psi_0 + \frac{\hbar^2}{2m\Delta x^2}\left(e^{-ik_L\Delta x} + (\psi_0-1)e^{ik_L\Delta x} - 2\psi_0 + \psi_1\right) - V_0 \psi_0 = 0$$

At the RHS boundary

$$E\psi_N + \frac{\hbar^2}{2m\Delta x^2}\left(\psi_{N-1} - 2\psi_N + "\psi"_{N+1}\right) - V_N \psi_N = 0$$

$$"\psi"_{N+1} = te^{ik_R(N+1)\Delta x} = e^{ik_R\Delta x}\psi_N \quad \text{(since } te^{ik_R N\Delta x} = \psi_N \text{ by continuity)}$$

$$E\psi_N + \frac{\hbar^2}{2m\Delta x^2}\left(\psi_{N-1} - 2\psi_N + e^{ik_R\Delta x}\psi_N\right) - V_N \psi_N = 0$$

$$\left[E\hat{I} - \begin{bmatrix} 2C+V_0-Ce^{ik_L\Delta x} & -C & 0 & \cdots \\ & \ddots & & \\ 0 & -C & 2C+V_i & -C & 0 & \cdots \\ & & & \ddots & \\ \cdots & 0 & -C & 2C+V_N-Ce^{ik_R\Delta x} \end{bmatrix}\right] \begin{bmatrix} \psi_0 \\ \psi_1 \\ \vdots \\ \vdots \\ \psi_N \end{bmatrix} = \begin{bmatrix} C(e^{ik_L\Delta x} - e^{-ik_L\Delta x}) \\ 0 \\ \vdots \\ \vdots \\ 0 \end{bmatrix}$$

$\left(C \equiv \frac{\hbar^2}{2m\Delta x^2}\right)$

An equivalent "Schrödinger" equation

This can be written

$$(E\hat{I} - \hat{H}')\vec{\Psi} = \vec{Q} \quad (\text{c.f. bound-state problems where "source" } Q=0)$$

Note, however, that the "Hamiltonian" $\hat{H}'$ is **NOT** Hermitian, due to the extra terms $(-Ce^{ik_y\Delta x}$ "self-energy") at the extremes of the main diagonal!

This means that, in general, the eigenvalues of H' are <u>not real!</u> As a result, the time dependence of Ψ $\left(\phi(t) = e^{-i\frac{E}{\hbar}t}\right)$ is not oscillatory but rather is exponentially decreasing.... There are no truly "<u>bound</u>" or "stationary" states that couple to propagating states! The incoming plane wave interacts w/ the scattering potential, and then after some time $\left(\sim \frac{\hbar}{\text{Imag}\{E\}}\right)$, leaves either to the left (reflected) or the right (transmitted)

$\longrightarrow$ Particle probability is <u>not</u> conserved!

Wavefunction and transmission coefficient

Transmission coefficient:

$$T = \frac{J_{trans}}{J_{inc}} = \frac{v_R |\psi_N|^2}{v_L |1|^2} \leftarrow |e^{ikx}|^2 \qquad \text{except } v \neq \frac{\hbar k}{m} \text{ in finite differences!}$$

$$J = \frac{\hbar}{m}\text{Im}\left\{\psi^*\frac{d\psi}{dx}\right\} = \frac{\hbar}{2mi}\left\{\psi^*\frac{d\psi}{dx} - \psi\frac{d\psi^*}{dx}\right\} = \frac{\hbar}{2mi}\left(\psi_i^*\frac{\psi_{i+1} - \psi_i}{\Delta x} - \psi_i\frac{\psi_{i+1}^* - \psi_i^*}{\Delta x}\right)$$

If $\psi = e^{ikx}$,

$$J = \frac{\hbar}{2mi\Delta x}\left(e^{-ikx_i}\left(e^{ik(x_i+\Delta x)} - e^{ikx_i}\right) - e^{ikx_i}\left(e^{-ik(x+\Delta x)} - e^{-ikx_i}\right)\right)$$

$$= \frac{\hbar}{2mi\Delta x}\left(e^{ik\Delta x} - 1 - (e^{-ik\Delta x} - 1)\right) = \frac{\hbar}{m\Delta x}\left(\frac{e^{ik\Delta x} - e^{-ik\Delta x}}{2i}\right) = \frac{\hbar}{m\Delta x}\sin k\Delta x$$

$$\sim \frac{\hbar k}{m} \text{ for } \Delta x \to 0$$

Dispersion relation

In continuous spatial dimension, $E = \frac{\hbar^2 k^2}{2m}$ so $\omega(k) = \frac{\hbar k^2}{2m}$

In finite differences, we can use $v = \frac{d\omega}{dk}$:

$$\omega(k) = \int v\, dk = \int \frac{\hbar}{m\Delta x} \sin k\Delta x\, dk = -\frac{\hbar}{m\Delta x^2} \cos k\Delta x + \text{const.} \longrightarrow \frac{\hbar}{m\Delta x^2}(1 - \cos k\Delta x)$$

$\cos k\Delta x \sim 1 - \frac{(k\Delta x)^2}{2}$ so $\omega(k) \sim \frac{\hbar}{m\Delta x^2}\left(1 - \left(1 - \frac{(k\Delta x)^2}{2}\right)\right) = \frac{\hbar k^2}{2m}$ ✓

Recipe

1. Choose E
2. Construct $\left[E\hat{I} - (\hat{H} + \hat{\Sigma}_L + \hat{\Sigma}_R)\right]$ and $\vec{Q}$
   using $k_{L/R} = \sqrt{\frac{2m(E - V_{L/R})}{\hbar^2}}$
3. Calculate $\vec{\Psi} = \left[E\hat{I} - (\hat{H} + \hat{\Sigma}_L + \hat{\Sigma}_R)\right]^{-1} \vec{Q}$
4. Calculate $T(E) = \frac{V_R}{V_L}|\Psi_N|^2$ using $V_{R/L} = \frac{\hbar}{m\Delta x}\sin k_{R/L}\Delta x$
5. Go to #1 and repeat w/ different E
6. Plot $T(E)$ vs. E.

Comparison between methods

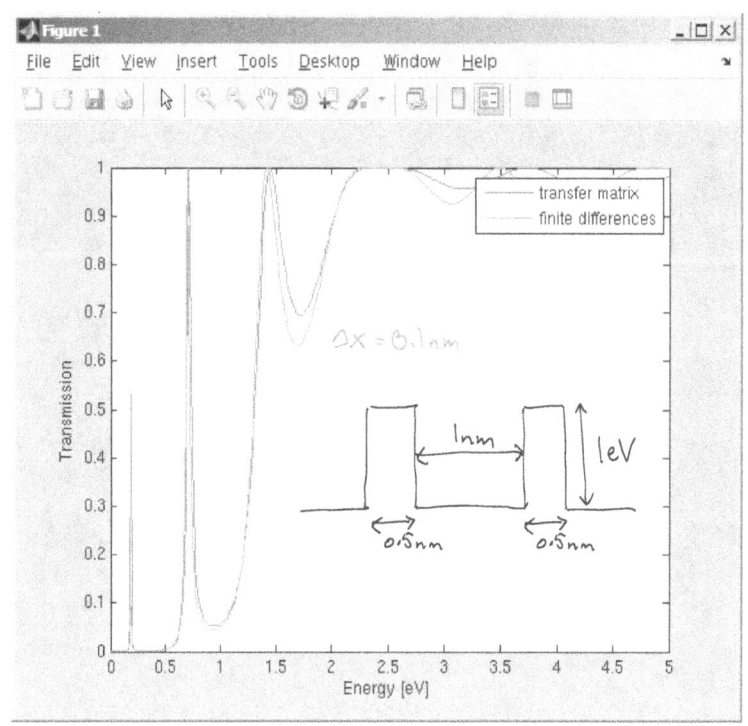

Current-voltage relations

Our model for a metal is a noninteracting "gas" of free electrons

Using ThB approx, charge current $I_{1D} = q\, n\, v_{avg} = q\dfrac{\sum_{i=1}^{N} v_i T_i}{L} = q \sum \dfrac{\hbar k}{m} T(k)/L$

(fundamental charge, electron density, Size of metal lead, "contributing states")

What are "contributing states"?

1. Available states $E = \dfrac{\hbar^2 k^2}{2m}$ "metal"

2. Occupied states $(T=0)$
   a. Minimization of energy
   b. Pauli exclusion principle

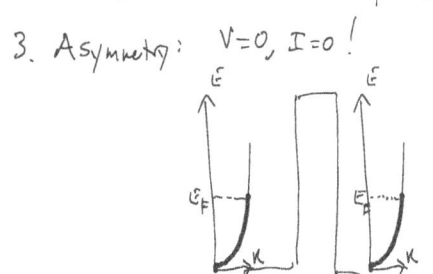

3. Asymmetry: $V=0, I=0$!

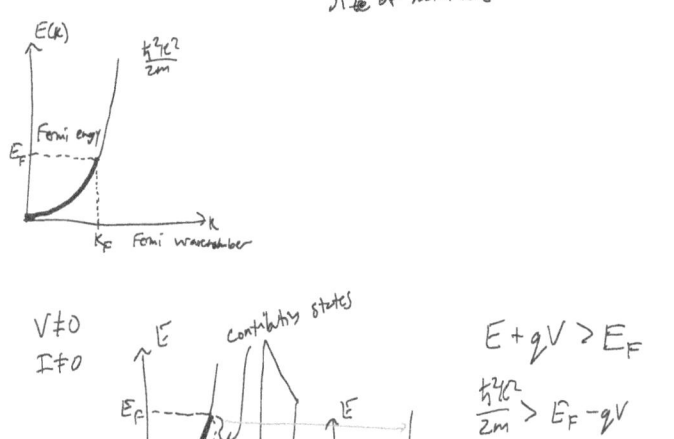

$E + qV > E_F$

$\dfrac{\hbar^2 k^2}{2m} > E_F - qV$

$k > k_{min} = \sqrt{\dfrac{2m(E_F - qV)}{\hbar^2}}$

Density of states

How much of k-space does each electron state take up?

Model infinite size of metal "leads" by using periodic boundary conditions:

$$\Psi(x) = \Psi(x+L)$$
$$e^{ikx} = e^{ik(x+L)} = e^{ikx} e^{ikL} \rightarrow kL = \text{integer multiples of } 2\pi$$

So successive states are separated by

$$\Delta k = \frac{2\pi}{L} \quad \text{(region of k-space occupied by 1 state)}$$

Our sum then becomes

$$I_{1D} = q \sum_{\text{contrib. states}} \frac{\hbar k}{m} T(k) \frac{\Delta k}{2\pi} \longrightarrow 2q \int_{k_{min}}^{k_F} \frac{\hbar k}{m} T(k,V) \frac{dk}{2\pi}$$

(with "spin degeneracy" arrow pointing to the 2)

For small V

In this regime, $k_{min} \approx k_F$ so contributing states all have $E \sim E_F$. Therefore, we can approximate $T(E,V) \approx T(E_F, V)$ (a constant). This allows us to take T out of the integral:

$$I_{1D} \approx 2q \frac{\hbar}{m} \frac{T}{2\pi} \int_{k_{min}}^{k_F} k \, dk = \frac{q\hbar T}{m\pi} \frac{k^2}{2} \Big|_{k_{min}}^{k_F}$$

$$= \frac{q\hbar T}{2\pi m} \left(k_F^2 - \frac{2m(E_F - qV)}{\hbar^2} \right)$$

$$= \frac{q\hbar T}{2\pi m} \left(\frac{2mqV}{\hbar^2} \right) \quad \left(\text{since } E_F = \frac{\hbar^2 k_F^2}{2m} \right)$$

$$= 2 \underbrace{\frac{q^2}{h}}_{\text{"quantum of conductance"}} T V \quad T \leq 1 \text{ so conductance is limited by } \frac{2q^2}{h}!$$

Quantum point contact

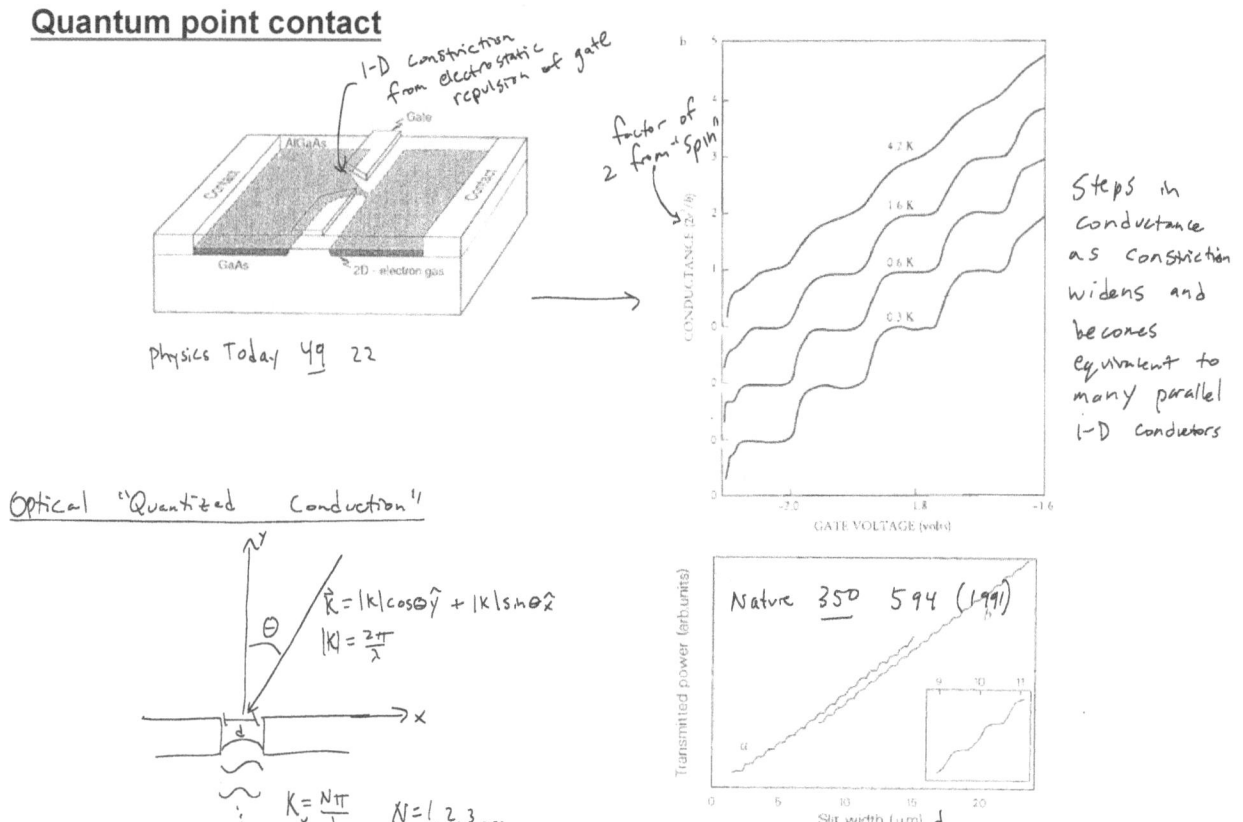

Physics Today **49** 22

1-D constriction from electrostatic repulsion of gate

factor of 2 from spin

Steps in conductance as constriction widens and becomes equivalent to many parallel 1-D conductors

Optical "Quantized Conduction"

$\vec{k} = |k|\cos\theta \hat{y} + |k|\sin\theta \hat{x}$

$|k| = \frac{2\pi}{\lambda}$

$k_x = \frac{N\pi}{d}$ $N = 1, 2, 3 \ldots$

Nature **350** 594 (1991)

Numerical results: single-barrier "tunnel junction"

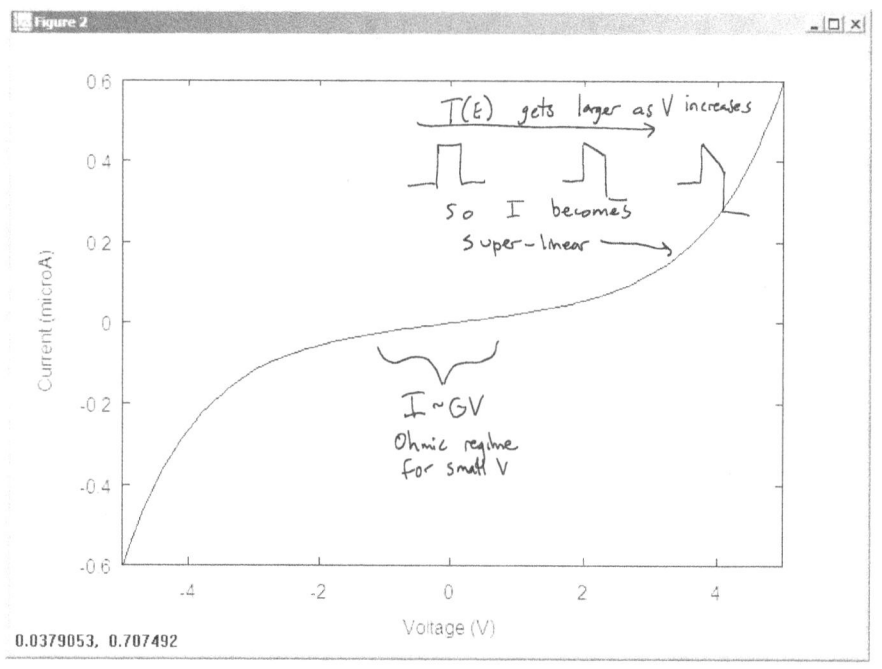

$T(E)$ gets larger as V increases

so I becomes super-linear

$I \sim GV$
Ohmic regime for small V

"Double-barrier resonant tunneling diode"

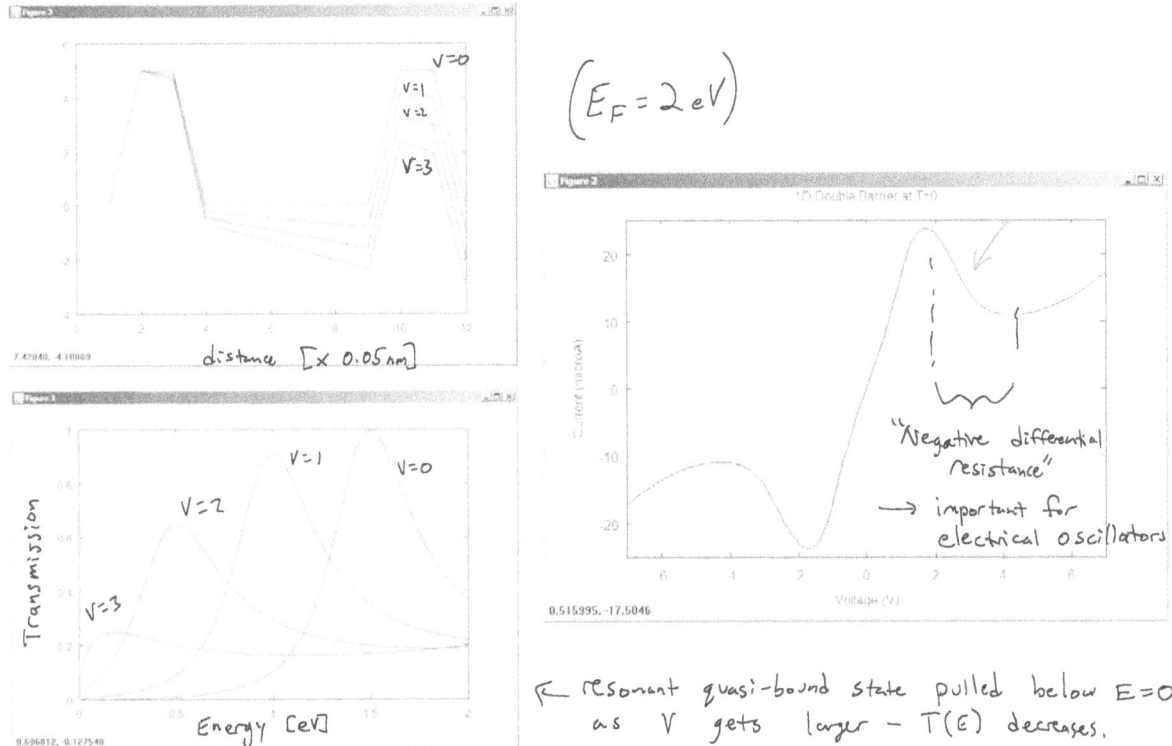

$(E_F = 2\,eV)$

"Negative differential resistance"
→ important for electrical oscillators

← resonant quasi-bound state pulled below E=0 as V gets larger — T(E) decreases.

Experimental evidence for negative differential resistance

Resonant tunneling through quantum wells at frequencies up to 2.5 THz

T. C. L. G. Sollner, W. D. Goodhue, P. E. Tannenwald, C. D. Parker, and D. D. Peck

Lincoln Laboratory, Massachusetts Institute of Technology, Lexington, Massachusetts 02173

Appl. Phys. Lett. 43 (6), 15 September 1983

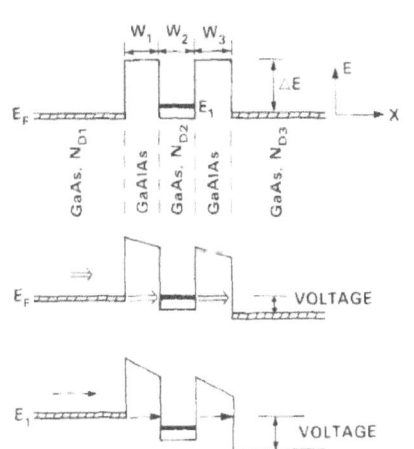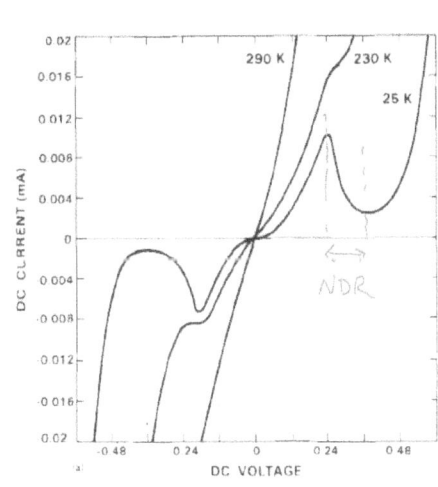

65

Hilbert space and Dirac notation

- An (in)finite-dimensional vector (or function) space of square-integrable vectors (or functions)
- Vectors (or functions) are denoted by "Kets" $|f\rangle$
- Inner product of $|f\rangle$ with $|g\rangle$ is obtained by taking Hermitian conjugate + multiplying $|f\rangle^\dagger |g\rangle$
- There is also a dual Hermitian-conjugate Hilbert space of "bras" $\langle g|$. Inner products can therefore be written
$$|f\rangle^\dagger |g\rangle = \langle f|g\rangle$$

This works for finite-dimensional vectors, or functions e.g. functions of x:
$$\langle f|g\rangle = \int_{-\infty}^{\infty} f^*(x) g(x)\, dx$$

Hilbert space and Dirac notation (cont)

- $\langle f|f\rangle$ is real and non-negative
- Orthonormal basis set $|f_n\rangle$ $n = 1, 2, 3, \ldots$
$$\langle f_n | f_m \rangle = \delta_{nm}$$
- basis set is complete if any vector (or function) in Hilbert space can be expressed as a linear combination of basis vectors (or functions)
$$|f\rangle = \sum_n c_n |f_n\rangle$$
Then, $\langle f_m | f \rangle = \sum_n c_n \langle f_m | f_n \rangle = \sum_n c_n \delta_{nm} = c_m$

So $c_n = \langle f_n | f \rangle$

Note:
$$|f\rangle = \sum_n |f_n\rangle \langle f_n | f \rangle = \left(\sum_n |f_n\rangle \langle f_n| \right) |f\rangle$$
$$\sum_n |f_n\rangle \langle f_n| = \hat{1} \quad (\text{identity})$$

Observables are Hermitian operators

- They have real eigenvalues and orthonormal eigenvectors (or eigenfunctions)
- Commute w/ vectors in Hilbert space

$$\langle f|Q|f\rangle = \langle f|Qf\rangle = (\langle f|Qf\rangle)^\dagger = \langle Q^\dagger f|f\rangle = \langle Qf|f\rangle$$

- The matrix elements of Q in the $|f_n\rangle$ basis are

$$\langle f_n|Q|f_m\rangle$$

Example: $Q = H \rightarrow \langle \psi_m|H|\psi_n\rangle = \langle \psi_m|E_n|\psi_n\rangle = E_n\langle \psi_m|\psi_n\rangle = E_n \delta_{mn}$

$$\begin{bmatrix} E_1 & 0 & 0 & 0 & \cdots \\ 0 & E_2 & 0 & & \\ 0 & 0 & E_3 & & \\ \vdots & \vdots & & \ddots & \end{bmatrix} \quad (\text{a } \underline{\text{diagonal}} \text{ matrix})$$

$\rightarrow$ The eigenfunctions/vectors of an operator "diagonalize" that operator!

Example

$$\hat{H} = \begin{bmatrix} 0 & 1 \\ 1 & 0 \end{bmatrix} \left(\text{in basis } |X\rangle = \begin{bmatrix} 1 \\ 0 \end{bmatrix} \text{ and } |Y\rangle = \begin{bmatrix} 0 \\ 1 \end{bmatrix} \right)$$

1. Show that $\langle n|\hat{H}|m\rangle$ is the nth row, m^{th} column element of H:

$H_{11} = \langle X|H|X\rangle = \begin{bmatrix} 1 & 0 \end{bmatrix}\begin{bmatrix} 0 & 1 \\ 1 & 0 \end{bmatrix}\begin{bmatrix} 1 \\ 0 \end{bmatrix} = \begin{bmatrix} 1 & 0 \end{bmatrix}\begin{bmatrix} 0 \\ 1 \end{bmatrix} = 0$, $\quad H_{12} = \langle X|H|Y\rangle = H_{21}^\dagger = 1$

$H_{21} = \langle Y|H|X\rangle = \begin{bmatrix} 0 & 1 \end{bmatrix}\begin{bmatrix} 0 & 1 \\ 1 & 0 \end{bmatrix}\begin{bmatrix} 1 \\ 0 \end{bmatrix} = \begin{bmatrix} 0 & 1 \end{bmatrix}\begin{bmatrix} 0 \\ 1 \end{bmatrix} = 1$, $\quad H_{22} = \begin{bmatrix} 0 & 1 \end{bmatrix}\begin{bmatrix} 0 & 1 \\ 1 & 0 \end{bmatrix}\begin{bmatrix} 0 \\ 1 \end{bmatrix} = \begin{bmatrix} 0 & 1 \end{bmatrix}\begin{bmatrix} 1 \\ 0 \end{bmatrix} = 0$

2. What are the eigenvalues of $\hat{H}$?

$\hat{H}|\psi\rangle = E|\psi\rangle$ (Schrödinger equation)

$(\hat{H} - E\hat{\mathbb{I}})|\psi\rangle = 0$

$\det \begin{bmatrix} -E & 1 \\ 1 & -E \end{bmatrix} = 0$

$E^2 - 1 = 0$
$E = \pm 1$
$E_+ = +1, \ E_- = -1$

Example (cont)

3. **What are the eigenvectors of $\hat{A}$?**

$(\hat{H} - E_+ \hat{I})|+\rangle = 0$

$\begin{bmatrix} -1 & 1 \\ 1 & -1 \end{bmatrix} \begin{bmatrix} a \\ b \end{bmatrix} = 0$

$|+\rangle = \frac{1}{\sqrt{2}} \begin{bmatrix} 1 \\ 1 \end{bmatrix}$

$(\hat{H} - E_- \hat{I})|-\rangle = 0$

$\begin{bmatrix} 1 & 1 \\ 1 & 1 \end{bmatrix} \begin{bmatrix} a \\ b \end{bmatrix} = 0$

$|-\rangle = \frac{1}{\sqrt{2}} \begin{bmatrix} 1 \\ -1 \end{bmatrix}$

4. **Show that $\sum_n |\psi_n\rangle\langle\psi_n| = \hat{I}$**

$|+\rangle\langle+| + |-\rangle\langle-|$

$\frac{1}{2}\left(\begin{bmatrix} 1 \\ 1 \end{bmatrix}\begin{bmatrix} 1 & 1 \end{bmatrix} + \begin{bmatrix} 1 \\ -1 \end{bmatrix}\begin{bmatrix} 1 & -1 \end{bmatrix}\right) = \frac{1}{2}\left(\begin{bmatrix} 1 & 1 \\ 1 & 1 \end{bmatrix} + \begin{bmatrix} 1 & -1 \\ -1 & 1 \end{bmatrix}\right) = \begin{bmatrix} 1 & 0 \\ 0 & 1 \end{bmatrix} = \hat{I}$ ✓

Numerical experiment

```
octave-3.2.2.exe:7:~\Desktop\PHYS401\code
> bra_ket(4)
A random complex-valued Hermitian matrix:
H =

   0.53427 + 0.00000i   1.55496 - 0.48363i   0.98593 + 0.41434i   1.04617 - 0.08737i
   1.55496 + 0.48363i   1.17180 + 0.00000i   0.69207 - 0.14460i   1.38173 - 0.10256i
   0.98593 - 0.41434i   0.69207 + 0.14460i   0.12701 + 0.00000i   1.28715 + 0.40134i
   1.04617 + 0.08737i   1.38173 - 0.10256i   1.28715 - 0.40134i   1.66651 + 0.00000i

Notice that the outer product (column vector times row vector) of eigenvectors is a matrix:
ans =

   0.41404 + 0.00000i  -0.23109 + 0.10564i  -0.25090 - 0.30816i   0.03233 + 0.13815i
  -0.23109 - 0.10564i   0.15593 + 0.00000i   0.06141 + 0.23601i   0.01720 - 0.08536i
  -0.25090 - 0.30816i   0.06141 - 0.23601i   0.38141 + 0.00000i  -0.12242 - 0.05965i
   0.03233 - 0.13815i   0.01720 + 0.08536i  -0.12242 + 0.05965i   0.04862 + 0.00000i

When we sum the outer products of eigenvectors, we get the identity to within machine precision
C =

   1.00000 + 0.00000i   0.00000 + 0.00000i   0.00000 - 0.00000i  -0.00000 - 0.00000i
   0.00000 - 0.00000i   1.00000 + 0.00000i   0.00000 - 0.00000i  -0.00000 + 0.00000i
   0.00000 + 0.00000i   0.00000 + 0.00000i   1.00000 + 0.00000i  -0.00000 - 0.00000i
  -0.00000 + 0.00000i  -0.00000 - 0.00000i  -0.00000 + 0.00000i   1.00000 + 0.00000i

the maximum error for the elements of this random matrix is:
1.3878e-016 + 2.0817e-016i
whereas machine precision on this computer is 2.2204e-016
octave-3.2.2.exe:8:~\Desktop\PHYS401\code
>
```

QM in higher spatial dimensions — 3D infinite cubical well

$$V(x,y,z) = \begin{cases} 0 & \begin{cases} 0<x<a_x \\ 0<y<a_y \\ 0<z<a_z \end{cases} \\ \infty & \text{otherwise} \end{cases}$$

Schrödinger Equation: $-\dfrac{\hbar^2}{2m}\left(\dfrac{\partial^2}{\partial x^2}+\dfrac{\partial^2}{\partial y^2}+\dfrac{\partial^2}{\partial z^2}\right)\psi(x,y,z) + V(x,y,z)\,\psi(x,y,z) = E\,\psi(x,y,z)$

Solution by separation of variables: $\psi(x,y,z) = \psi_x(x)\,\psi_{yz}(y,z)$

$$\dfrac{-\dfrac{\hbar^2}{2m}\left[\left(\dfrac{\partial^2}{\partial x^2}\psi_x\right)\psi_{yz} + \psi_x\left(\dfrac{\partial^2}{\partial y^2}+\dfrac{\partial^2}{\partial z^2}\right)\psi_{yz}\right]}{\psi_x\psi_{yz}} = \dfrac{E\,\psi_x\psi_{yz}}{\psi_x\psi_{yz}}$$

$$\dfrac{-\dfrac{\hbar^2}{2m}\dfrac{\partial^2}{\partial x^2}\psi_x}{\psi_x} - E = \dfrac{\dfrac{\hbar^2}{2m}\left(\dfrac{\partial^2}{\partial y^2}+\dfrac{\partial^2}{\partial z^2}\right)\psi_{yz}}{\psi_{yz}} = -E_{yz}$$

① $\boxed{-\dfrac{\hbar^2}{2m}\dfrac{d^2}{dx^2}\psi_x = (E-E_{yz})\psi_x = E_x\psi_x}$ 	 $-\dfrac{\hbar^2}{2m}\left(\dfrac{\partial^2}{\partial y^2}+\dfrac{\partial^2}{\partial z^2}\right)\psi_{yz} = E_{yz}\psi_{yz}$

Separation of Variables (Again!)

$\psi_{yz}(y,z) = \psi_y(y)\,\psi_z(z)$

$$\dfrac{-\dfrac{\hbar^2}{2m}\left[\psi_z\dfrac{\partial^2}{\partial y^2}\psi_y + \psi_y\dfrac{\partial^2}{\partial z^2}\psi_z\right]}{\psi_y\psi_z} = \dfrac{E_{yz}\psi_y\psi_z}{\psi_y\psi_z}$$

$$\dfrac{-\dfrac{\hbar^2}{2m}\dfrac{\partial^2}{\partial y^2}\psi_y}{\psi_y} - E_{yz} = \dfrac{\dfrac{\hbar^2}{2m}\dfrac{\partial^2}{\partial z^2}\psi_z}{\psi_z} = -E_z$$

② $\boxed{-\dfrac{\hbar^2}{2m}\dfrac{d^2}{dy^2}\psi_y = (E_{yz}-E_z)\psi_y = E_y\psi_y}$ ③ $\boxed{-\dfrac{\hbar^2}{2m}\dfrac{d^2}{dz^2}\psi_z = E_z\psi_z}$

Note: $E_x + E_y + E_z = (E - E_{yz}) + (E_{yz} - E_z) + E_z = E\ !$

Full solution

Solve each equation independently, Then sum eigenvalues.

Apply B.C.'s
$$\psi(0,y,z) = \psi(a_x,y,z) = 0 \rightarrow \psi_x(0) = \psi(a_x) = 0$$
$$\psi(x,0,z) = \psi(x,a_y,z) = 0 \rightarrow \psi_y(0) = \psi(a_y) = 0$$
$$\psi(x,y,0) = \psi(x,y,a_z) = 0 \rightarrow \psi_z(0) = \psi(a_z) = 0$$

This yields familiar solutions $\sqrt{\frac{2}{a_x}} \sin k_x x$, $\sqrt{\frac{2}{a_y}} \sin k_y y$ and $\sqrt{\frac{2}{a_z}} \sin k_z z$ with eigenvalues:

$E_x = \frac{\hbar^2}{2m} k_x^2 = \frac{\hbar^2 \pi^2 n_x^2}{2m a_x^2}$ $n_x = 1,2,3,...$

$E_y = \frac{\hbar^2}{2m} k_y^2 = \frac{\hbar^2 \pi^2 n_y^2}{2m a_y^2}$ $n_y = 1,2,3,...$ so $E = E_x + E_y + E_z = \frac{\hbar^2}{2m}\left(\frac{n_x^2}{a_x^2} + \frac{n_y^2}{a_y^2} + \frac{n_z^2}{a_z^2}\right)$

$E_z = \frac{\hbar^2}{2m} k_z^2 = \frac{\hbar^2 \pi^2 n_z^2}{2m a_z^2}$ $n_z = 1,2,3,...$

Full eigenfunction given by $\Psi(x,y,z,t) = \psi_{n_x}(x)\, \psi_{n_y}(y)\, \psi_{n_z}(z)\, e^{-i\frac{E}{\hbar}t}$

Total Energies

(for $a_x = a_y = a_z = a$)

n_x	n_y	n_z	$E\left(\frac{\hbar^2}{2ma^2}\right)$	degeneracy
1	1	1	3	1
1	1	2		
1	2	1	6	3
2	1	1		
1	2	2		
2	1	2	9	3
2	2	1		
3	1	1		
1	3	1	11	3
1	1	3		
2	2	2	12	1
1	2	3		
1	3	2		
2	1	3	14	6
2	3	1		
3	1	2		
3	2	1		

These degenercies are caused by symmetry (equivalence of x, y, and z dimensions for $a_x = a_y = a_z$. If, instead, $a_x \neq a_y \neq a_z$, the degenercies will be broken, except for "accidental" degenercies when
$$\frac{n_x^2}{a_x^2} + \frac{n_y^2}{a_y^2} + \frac{n_z^2}{a_z^2} = \frac{n_x'^2}{a_x^2} + \frac{n_y'^2}{a_y^2} + \frac{n_z'^2}{a_z^2}$$

Finite differences in 3D

$$-\frac{\hbar^2}{2m}\nabla^2 \psi_{ijk} + V_{ijk}\psi_{ijk} = E\psi_{ijk}$$

$\begin{pmatrix} \text{integer} \\ i: \text{index along } \hat{x} \\ j: \text{index along } \hat{y} \\ k: \text{index along } \hat{z} \end{pmatrix}$ Volume element: at location $(x,y,z) = (i\Delta x, j\Delta y, k\Delta z)$

System of linear equations:

$$t_x(\psi_{i+1,j,k} - 2\psi_{ijk} + \psi_{i-1,j,k}) + t_y(\psi_{i,j+1,k} - 2\psi_{ijk} + \psi_{i,j-1,k}) + t_z(\psi_{i,j,k+1} - 2\psi_{ijk} + \psi_{i,j,k-1}) + V_{ijk}\psi_{ijk} = E\psi_{ijk}$$

$$\left(t_x \equiv -\frac{\hbar^2}{2m\Delta x^2}, \quad t_y \equiv -\frac{\hbar^2}{2m\Delta y^2}, \quad t_z \equiv -\frac{\hbar^2}{2m\Delta z^2}\right)$$

Example: 2×2×2 discretization of infinite cubical well

Application of finite differences results in 8×8 matrix eigenvalue problem

Matrix Hamiltonian

$\left(\text{just } H = -\frac{\hbar^2}{2m}\nabla^2 \text{ Kinetic energy term}\right)$

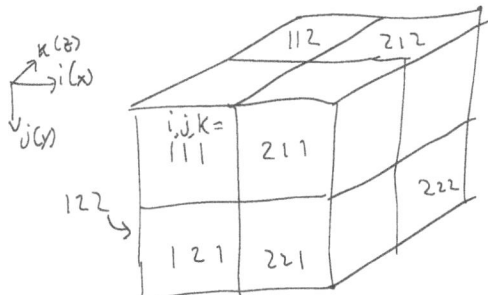

	ijk=111	211	121	221	112	212	122	222
111	$-2(t_x+t_y+t_z)$	t_x	t_y	0	t_z	0	0	0
211	t_x	$-2(t_x+t_y+t_z)$	0	t_y	0	t_z	0	0
121	t_y	0	$-2(t_x+t_y+t_z)$	t_x	0	0	t_z	0
221	0	t_y	t_x	$-2(t_x+t_y+t_z)$	0	0	0	t_z
112	t_z	0	0	0	$-2(t_x+t_y+t_z)$	t_x	t_y	0
212	0	t_z	0	0	t_x	$-2(t_x+t_y+t_z)$	0	t_y
122	0	0	t_z	0	t_y	0	$-2(t_x+t_y+t_z)$	t_x
222	0	0	0	t_z	0	t_y	t_x	$-2(t_x+t_y+t_z)$

BLOCK tridiagonal!

→ This symmetry extends to arbitrary matrix sizes!

(in a general 3-d problem, V_{ijk} will appear along the diagonal)

Larger Hamiltonian

(Ex. $3 \times 3 \times 3$)

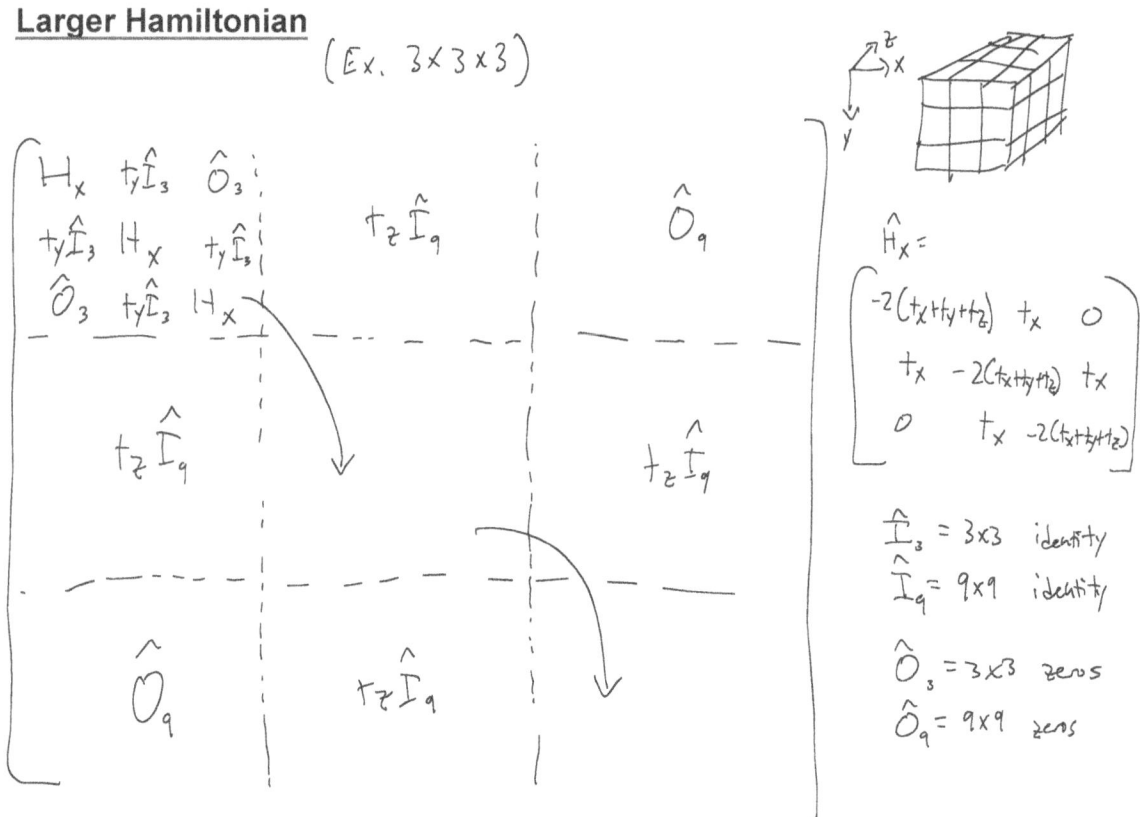

Numerical Results

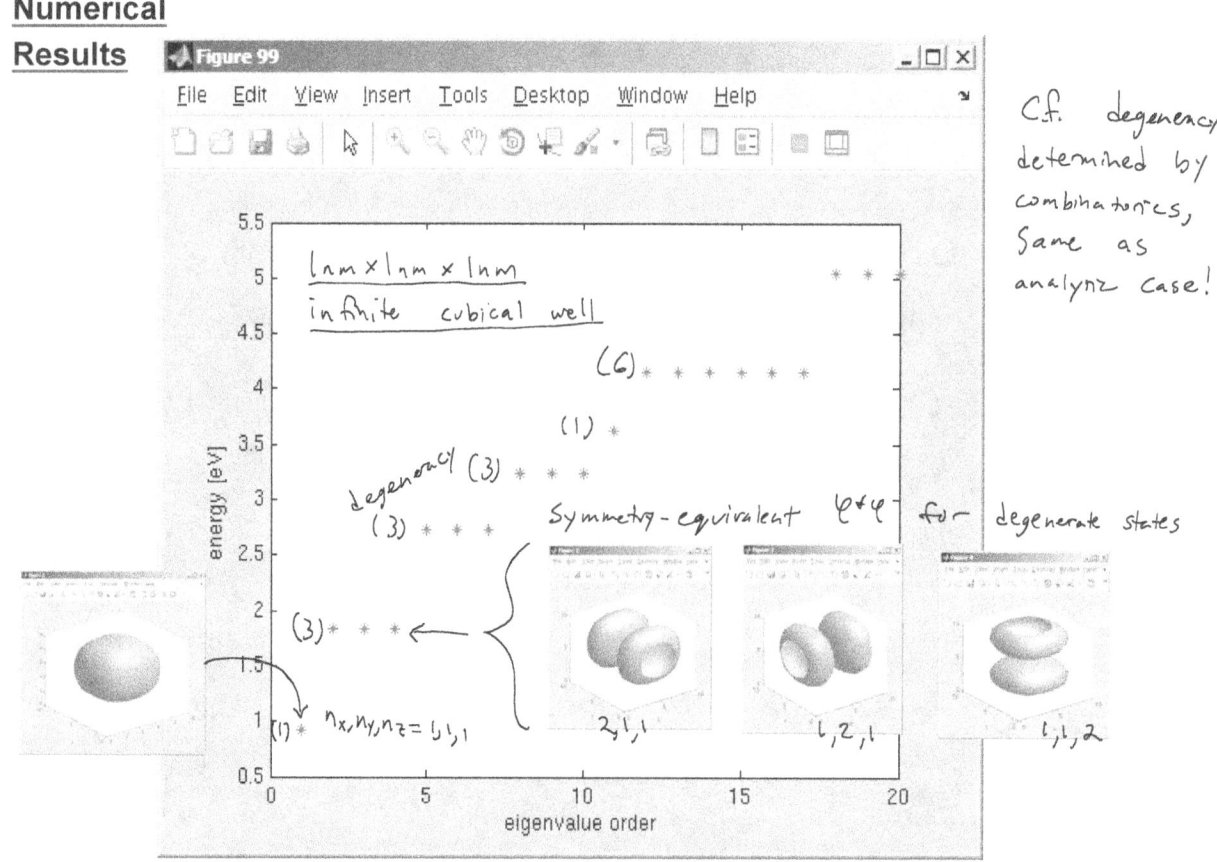

Cf. degeneracy determined by combinatorics, same as analytic case!

$\psi^*\psi$ for degenerate states

Constructing 3D matrix Hamiltonians

Kronecker (tensor) matrix product "$\otimes$":

Example:
$$\begin{bmatrix} 1 & 2 \\ 3 & 4 \end{bmatrix} \otimes \begin{bmatrix} a & b \\ c & d \end{bmatrix} = \begin{bmatrix} 1\begin{bmatrix} a & b \\ c & d \end{bmatrix} & 2\begin{bmatrix} a & b \\ c & d \end{bmatrix} \\ 3\begin{bmatrix} a & b \\ c & d \end{bmatrix} & 4\begin{bmatrix} a & b \\ c & d \end{bmatrix} \end{bmatrix}$$

(2×2)   (2×2)   (4×4)

A 3×3×3 matrix Hamiltonian

$$I_x \otimes I_y \otimes H_z$$
$$+ I_x \otimes H_y \otimes I_z$$
$$+ H_x \otimes I_y \otimes I_z$$

In Matlab/Octave: $A \otimes B \rightarrow \text{kron}(A,B)$
$|\psi|^2_{3D} = \text{reshape}(|\psi|^2_{1D}, N_x, N_y, N_z)$
$\text{isosurface}(|\psi|^2_{3D}, \text{density})$

Broken symmetry

non-cubical infinite box

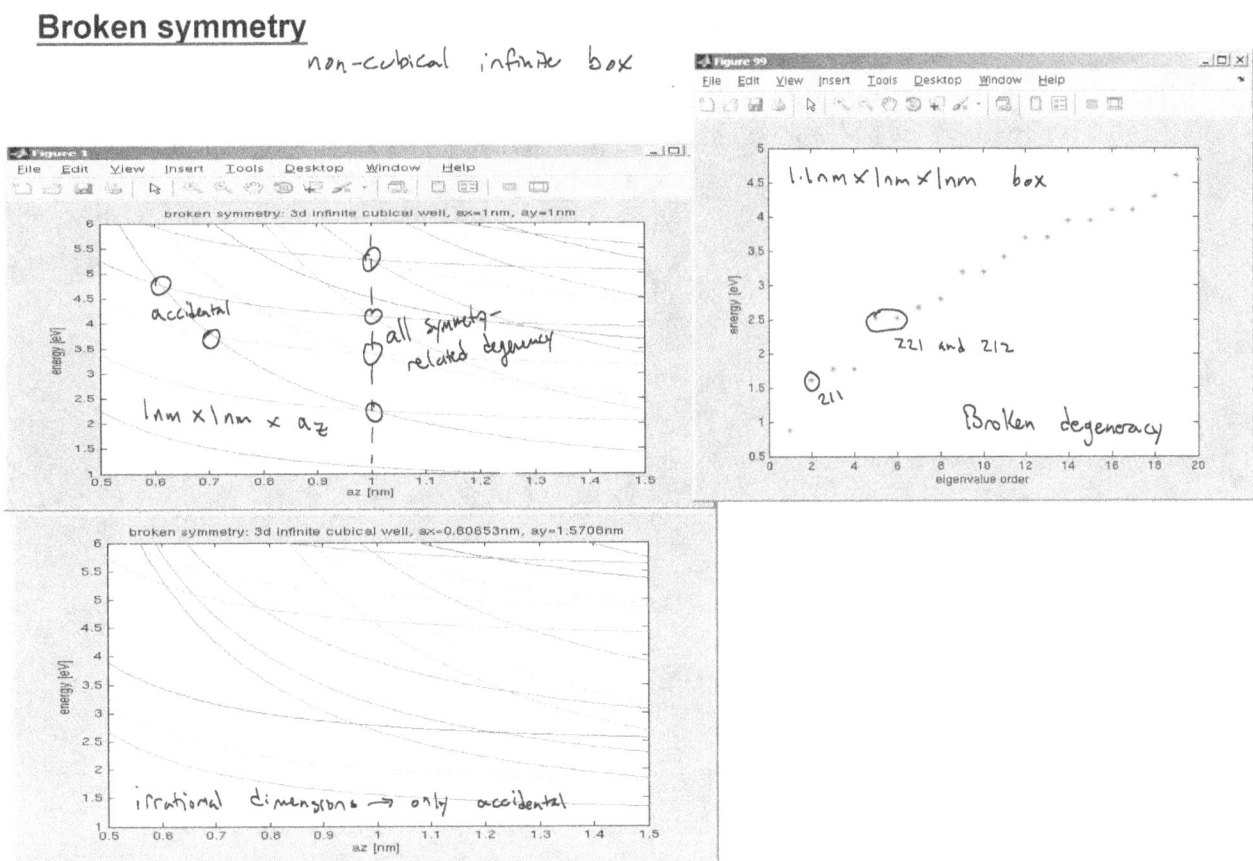

3D isotropic harmonic oscillator

$$-\frac{\hbar^2}{2m}\nabla^2\psi + \frac{1}{2}m\omega^2 r^2\psi = E\psi, \quad r=\sqrt{x^2+y^2+z^2} \quad \text{so } V(x) = \frac{1}{2}m\omega^2(x^2+y^2+z^2)$$

This is a separable potential, giving

$$E = E_x + E_y + E_z = \hbar\omega(n_x+\tfrac{1}{2}) + \hbar\omega(n_y+\tfrac{1}{2}) + \hbar\omega(n_z+\tfrac{1}{2}) = \hbar\omega\left(n_x+n_y+n_z+\tfrac{3}{2}\right), \quad n_x, n_y, n_z = 0,1,2,\dots$$

n_x	n_y	n_z	$E\ (\hbar\omega)$	degeneracy
0	0	0	3/2	1
1	0	0		
0	1	0	5/2	3
0	0	1		
2	0	0		
0	2	0		
0	0	2	7/2	6
1	1	0		
1	0	1		
0	1	1		

Finite difference calculation of 3D HO spectrum

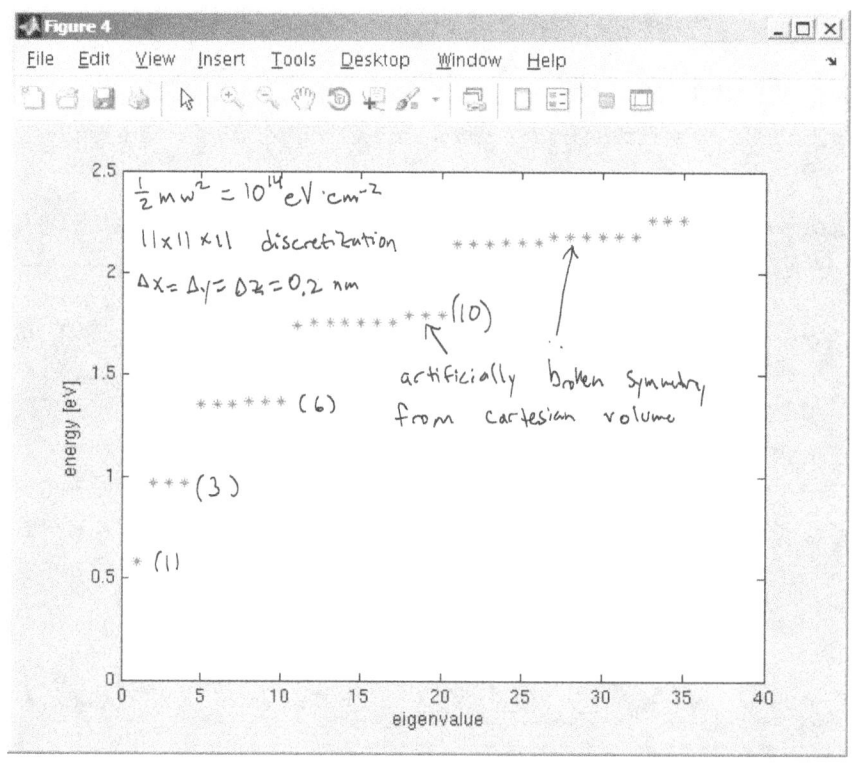

$\frac{1}{2}m\omega^2 = 10^{14}\ \text{eV}\cdot\text{cm}^{-2}$

$11\times 11\times 11$ discretization

$\Delta x = \Delta y = \Delta z = 0.2$ nm

artificially broken symmetry from cartesian volume

Classical waves in 1-D (review)

[Diagram: string under tension T between two walls, displacement y, segment from x to $x+\Delta x$, mass density ρ]

$F = ma \longrightarrow T \dfrac{\partial y(x+\Delta x)}{\partial x} - T \dfrac{\partial y(x)}{\partial x} = (\rho \Delta x)\dfrac{\partial^2 y}{\partial t^2}$

$\sim T\left(\dfrac{\partial}{\partial x}\left(y(x) + \Delta x \dfrac{\partial y(x)}{\partial x}\right)\right) - T\dfrac{\partial y(x)}{\partial x} = \rho \Delta x \dfrac{\partial^2 y}{\partial t^2}$

$$\dfrac{\partial^2 y}{\partial x^2} = \dfrac{\rho}{T}\dfrac{\partial^2 y}{\partial t^2} = \dfrac{1}{c^2}\dfrac{\partial^2 y}{\partial t^2} \qquad \left(\dfrac{1}{c^2} = \dfrac{\rho}{T}\right)$$

Classical waves on a circle

[Diagram: circle of radius r, angle ϕ, arc element Δx]

$\Delta x = r\Delta\phi \to dx = r\,d\phi$

$\dfrac{1}{c^2}\dfrac{\partial^2 y}{\partial t^2} = \dfrac{\partial^2 y}{r^2 \partial \phi^2}$

$y(\phi, t) = T(t)\,\psi(\phi)$
$\qquad \dfrac{\frac{1}{c^2}\ddot{T}\psi}{T\psi} = \dfrac{\frac{1}{r^2}T\psi''}{T\psi} = -k^2$

$\ddot{T} = -k^2 c^2 T \to T = A_\pm e^{\pm ikct} \qquad \psi'' = -k^2 r^2 \psi \to \psi = B e^{\pm ikr\phi}$

Apply periodic B.C.'s:

$\psi(\phi) = \psi(\phi + 2\pi) \qquad e^{ikr\phi} = e^{ikr(\phi+2\pi)} = e^{ikr\phi}\underbrace{e^{ikr2\pi}}_{=1}$

So $kr = n$, $n = 0, \pm 1, \pm 2, \ldots$

$\psi \propto e^{in\phi}$

Classical waves on the surface of a sphere

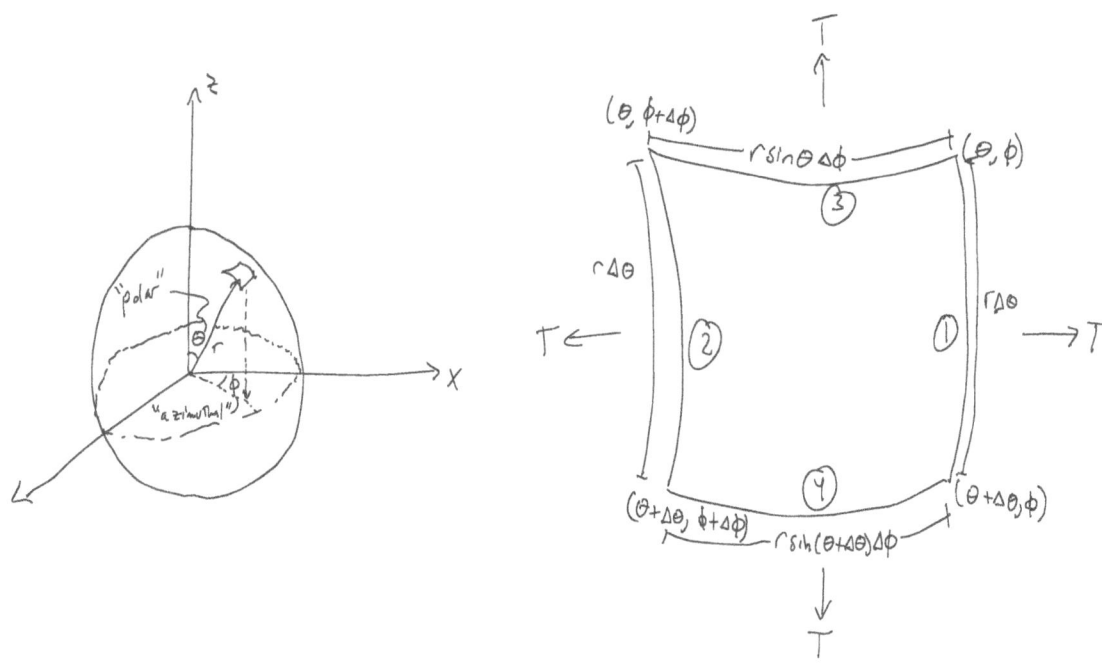

Newton's 2nd law

$F = ma$

$$② - ① + ④ - ③ = \rho (r\Delta\theta \cdot r\sin\theta \Delta\phi) \frac{\partial^2 f(\theta,\phi,t)}{\partial t^2}$$

①: $T r\Delta\theta \dfrac{\partial f(\theta,\phi,t)}{r\sin\theta \partial \phi}$

②: $T r\Delta\theta \dfrac{\partial f(\theta, \phi+\Delta\phi, t)}{r\sin\theta \partial\phi} \sim \dfrac{T\Delta\theta}{\sin\theta} \dfrac{\partial}{\partial\phi}\left(f(\theta,\phi,t) + \dfrac{\partial f(\theta,\phi,t)}{\partial\phi}\Delta\phi\right)$

③: $T r\sin\theta \Delta\phi \dfrac{\partial f(\theta,\phi,t)}{r\partial\theta}$

④: $T r\sin(\theta+\Delta\theta)\Delta\phi \dfrac{\partial f(\theta+\Delta\theta,\phi,t)}{r\partial\theta} \sim T\Delta\phi \left(\sin\theta + \Delta\theta \dfrac{d\sin\theta}{d\theta}\right)\left(\dfrac{\partial}{\partial\theta}\left(f(\theta,\phi,t) + \Delta\theta \dfrac{\partial f(\theta,\phi,t)}{\partial\theta}\right)\right)$

$\sim T\Delta\phi \left(\sin\theta \dfrac{\partial f}{\partial\theta} + \Delta\theta \dfrac{\partial \sin\theta}{\partial\theta} \dfrac{\partial f}{\partial\theta} + \Delta\theta \sin\theta \dfrac{\partial^2 f}{\partial\theta^2}\right)$

$= T\Delta\phi \left(\sin\theta \dfrac{\partial f}{\partial\theta} + \Delta\theta \dfrac{\partial}{\partial\theta}\left(\sin\theta \dfrac{\partial f}{\partial\theta}\right)\right)$

Equation of motion

$$\textcircled{2} - \textcircled{1} + \textcircled{4} - \textcircled{3} = \rho(r\Delta\theta \cdot r\sin\theta \Delta\phi) \frac{\partial^2 f(\theta,\phi,t)}{\partial t^2}$$

$$\frac{T}{\sin\theta} \frac{\partial^2 f}{\partial \phi^2} \Delta\theta\Delta\phi + T\frac{\partial}{\partial \theta}\left(\sin\theta \frac{\partial f}{\partial \theta}\right)\Delta\theta\Delta\phi = \rho(r\Delta\theta \cdot r\sin\theta\Delta\phi)\frac{\partial^2 f(\theta,\phi,t)}{\partial t^2}$$

$$\frac{1}{r^2\sin^2\theta}\frac{\partial^2 f}{\partial \phi^2} + \frac{1}{r^2\sin\theta}\frac{\partial}{\partial \theta}\left(\sin\theta \frac{\partial f}{\partial \theta}\right) = \frac{1}{c^2}\frac{\partial^2 f}{\partial t^2} \qquad \left(c^2 = \frac{T}{\rho}\right)$$

$$\nabla^2 f = \frac{1}{c^2}\frac{\partial^2 f}{\partial t^2}$$

This is a wave equation where $\nabla^2 \equiv \frac{1}{r^2\sin^2\theta}\frac{\partial^2}{\partial \phi^2} + \frac{1}{r^2\sin\theta}\frac{\partial}{\partial \theta}\left[\sin\theta \frac{\partial}{\partial \theta}\right]$

Separation of Variables

$$\frac{1}{\sin^2\theta}\frac{\partial^2 f}{\partial \phi^2} + \frac{1}{\sin\theta}\frac{\partial}{\partial \theta}\left(\sin\theta\frac{\partial f}{\partial \theta}\right) = \frac{r^2}{c^2}\frac{\partial^2 f}{\partial t^2}$$

$f(\theta,\phi,t) = Y(\theta,\phi) T(t)$

$$\frac{\frac{1}{\sin^2\theta}\frac{\partial^2 Y}{\partial \phi^2} + \frac{1}{\sin\theta}\frac{\partial}{\partial \theta}\left(\sin\theta\frac{\partial Y}{\partial \theta}\right)}{Y} = \frac{\frac{r^2}{c^2}\frac{\partial^2 T}{\partial t^2}}{T} = -\lambda \quad \text{(unitless)}$$

$T(t): \quad \frac{\partial^2 T}{\partial t^2} = -\lambda \frac{c^2}{r^2} T \quad \rightarrow \quad T(t) = A_{\pm} e^{\pm i \frac{c}{r}\sqrt{\lambda} t}$

θ and ϕ:
$$\frac{1}{\sin^2\theta}\frac{\partial^2 Y}{\partial \phi^2} + \frac{1}{\sin\theta}\frac{\partial}{\partial \theta}\left(\sin\theta\frac{\partial Y}{\partial \theta}\right) = -\lambda Y$$

$Y(\theta,\phi) = F_\phi(\phi) F_\theta(\theta):$
$$\frac{\frac{\partial^2}{\partial \phi^2}F_\phi}{F_\phi} = \frac{-\sin\theta\left(\frac{1}{\sin\theta}\frac{\partial}{\partial \theta}\left(\sin\theta \frac{\partial F_\theta}{\partial \theta}\right) + \lambda F_\theta\right)}{F_\theta} = -m^2$$

Azimuthal Equation

$$\frac{\partial^2}{\partial \phi^2} F_\phi = -m^2 F_\phi \quad \rightarrow \quad F_\phi(\phi) = A e^{im\phi}$$

Apply periodic boundary conditions: $F_\phi(\phi) = F_\phi(\phi + 2\pi)$

$$A e^{im\phi} = A e^{im(\phi + 2\pi)} = A e^{im\phi} e^{im 2\pi}$$

So $m = 0, \pm 1, \pm 2 \ldots$

Polar equation

$$-\sin^2\theta \left(\frac{1}{\sin\theta} \frac{\partial}{\partial \theta}\left(\sin\theta \frac{\partial F_\theta}{\partial \theta} \right) + \lambda F_\theta \right) = -m^2 F_\theta$$

If $m = 0$

$$\frac{1}{\sin\theta} \frac{d}{d\theta}\left(\sin\theta \frac{dF_\theta}{d\theta} \right) = -\lambda F_\theta$$

If $\lambda = 0$, $F_\theta(\theta) = $ constant

If $\lambda \neq 0$ $F_\theta(\theta) \stackrel{?}{=} \cos\theta$

$$\frac{1}{\sin\theta} \frac{d}{d\theta}\left(\sin\theta \cdot (-\sin\theta) \right) \stackrel{?}{=} \lambda \cos\theta$$

$$-\frac{1}{\sin\theta} \cdot 2\sin\theta \cos\theta = -\lambda \cos\theta \quad \rightarrow \quad \text{only if } \lambda = 2!$$

What about higher powers?

$F_\theta \stackrel{?}{=} \cos^2\theta$?

$$\frac{1}{\sin\theta}\frac{d}{d\theta}\left(\sin\theta \frac{d}{d\theta}\cos^2\theta\right) \stackrel{?}{=} -\lambda \cos^2\theta$$

$$-\frac{1}{\sin\theta}\frac{d}{d\theta}(2\cos\theta \sin^2\theta)$$

$$-\frac{1}{\sin\theta}(-2\sin^3\theta + 2\cos\theta \cdot 2\sin\theta\cos\theta) = 2\sin^2\theta - 4\cos^2\theta$$

$$= 2\sin^2\theta + 2\cos^2\theta - 6\cos^2\theta$$

$$= 2 - 6\cos^2\theta \neq -\lambda\cos^2\theta$$

But $F_\theta(\theta) = \cos^2\theta - \frac{1}{3}$ works, with $\lambda = 6$

So, for $m=0$, $\lambda = 0, 2, 6, (12), \ldots$ $\lambda_\ell = \ell(\ell+1)$, $\ell = 0, 1, 2, \ldots$

Example for m≠0

$\ell = 1$ ($\lambda = 2$)

$$-\sin^2\theta\left(2F_\theta + \frac{1}{\sin\theta}\frac{d}{d\theta}\left(\sin\theta \frac{d}{d\theta}F_\theta\right)\right) = -m^2 F_\theta$$

$F_\theta \stackrel{?}{=} \sin\theta$?:

$$-\sin^2\theta\left(2\sin\theta + \frac{1}{\sin\theta}\frac{d}{d\theta}\left(\sin\theta \frac{d}{d\theta}\sin\theta\right)\right) = -m^2 \sin\theta$$

$$-\sin^2\theta\left(2\sin\theta + \frac{1}{\sin\theta}\frac{d}{d\theta}(\sin\theta\cos\theta)\right) =$$

$$-\sin^2\theta\left(2\sin\theta + \frac{1}{\sin\theta}(\cos^2\theta - \sin^2\theta)\right) = -m^2\sin\theta$$

$$-\sin\theta\left(2\sin^2\theta + \cos^2\theta - \sin^2\theta\right)$$

$$-\sin\theta = -m^2\sin\theta$$

$$\Rightarrow m = \pm 1$$

In general, $m = -\ell, -\ell+1, \ldots, \ell-1, +\ell$

Solution to angular equation

$$Y_{\ell m}(\theta,\phi) = F_\theta^{\ell m}(\theta) F_\phi^m(\phi) \qquad \text{"spherical harmonics"}$$

$\ell = 0, 1, 2, 3$; $m = -3, -2, -1, 0, +1, +2, +3$

Extension to 3D

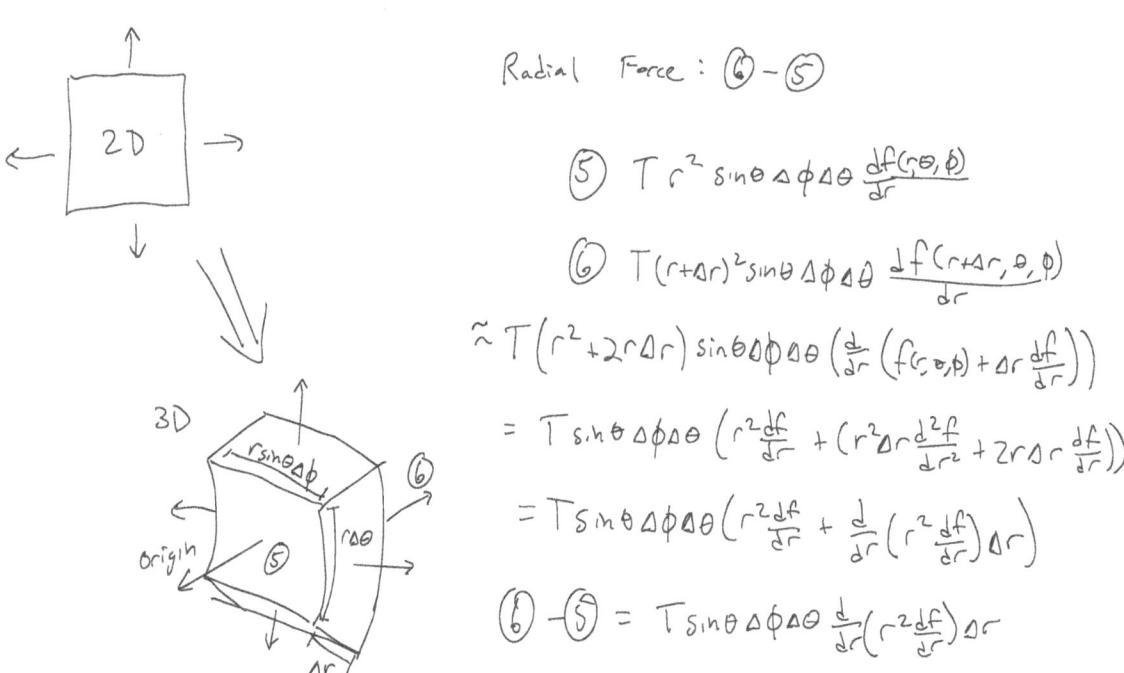

Radial Force: ⑥ − ⑤

⑤ $\;T r^2 \sin\theta\, \Delta\phi\, \Delta\theta\, \dfrac{df(r,\theta,\phi)}{dr}$

⑥ $\;T(r+\Delta r)^2 \sin\theta\, \Delta\phi\, \Delta\theta\, \dfrac{df(r+\Delta r,\theta,\phi)}{dr}$

$\;\approx T(r^2 + 2r\Delta r)\sin\theta\, \Delta\phi\, \Delta\theta\, \left(\dfrac{d}{dr}\left(f(r,\theta,\phi) + \Delta r \dfrac{df}{dr}\right)\right)$

$\;= T\sin\theta\, \Delta\phi\, \Delta\theta\, \left(r^2 \dfrac{df}{dr} + \left(r^2 \Delta r \dfrac{d^2 f}{dr^2} + 2r\Delta r \dfrac{df}{dr}\right)\right)$

$\;= T\sin\theta\, \Delta\phi\, \Delta\theta\, \left(r^2 \dfrac{df}{dr} + \dfrac{d}{dr}\left(r^2 \dfrac{df}{dr}\right)\Delta r\right)$

⑥ − ⑤ $= T\sin\theta\, \Delta\phi\, \Delta\theta\, \dfrac{d}{dr}\left(r^2 \dfrac{df}{dr}\right)\Delta r$

Newton's 2nd Law in 3D

$$\frac{T}{\sin\theta}\frac{\partial^2 f}{\partial \phi^2}\Delta\phi\Delta\theta\Delta r + T\frac{\partial}{\partial \theta}\left(\sin\theta \frac{\partial f}{\partial \theta}\right)\Delta\phi\Delta\theta\Delta r + T\sin\theta\Delta\theta\Delta\phi\Delta r \frac{\partial}{\partial r}\left(r^2 \frac{\partial f}{\partial r}\right)$$

$$= \rho r^2 \sin\theta \Delta\theta \Delta\phi \Delta r \frac{\partial^2 f}{\partial t^2}$$

$$\underbrace{\left[\frac{1}{r^2}\frac{\partial}{\partial r}r^2\frac{\partial}{\partial r} + \frac{1}{r^2\sin\theta}\frac{\partial}{\partial \theta}\left(\sin\theta \frac{\partial}{\partial \theta}\right) + \frac{1}{r^2 \sin^2\theta}\frac{\partial^2}{\partial \phi^2}\right]}_{\nabla^2_{3D} \text{ in spherical curvilinear coords}} f = \frac{1}{c^2}\frac{\partial^2 f}{\partial t^2} \qquad \left(c^2 = \frac{T}{\rho}\right)$$

Time-independent Schrödinger Eq in 3D spherical coordinates

1-D: $\quad -\frac{\hbar^2}{2m}\frac{d^2}{dx^2}\psi(x) + V(x)\psi(x) = E\psi(x)$

3-D: $\quad -\frac{\hbar^2}{2m}\nabla^2_{3D}\psi(r,\theta,\phi) + V(r)\psi(r,\theta,\phi) = E\psi(r,\theta,\phi)$

Separation of variables: $\quad \psi(r,\theta,\phi) = R(r)Y(\theta,\phi)$

$$-\frac{\hbar^2}{2m}\left[\left(\frac{1}{r^2}\frac{\partial}{\partial r}r^2\frac{\partial}{\partial r}R\right)Y + \left(\nabla^2_{2D}Y\right)R\right] + V(r)RY = ERY$$

$$\frac{\frac{\partial}{\partial r}\left(r^2\frac{\partial}{\partial r}R\right) - \frac{2mr^2}{\hbar^2}(V(r)-E)R}{R} = \underbrace{\frac{-\left(\frac{1}{\sin\theta}\frac{\partial}{\partial \theta}\sin\theta\frac{\partial Y}{\partial \theta} + \frac{1}{\sin^2\theta}\frac{\partial^2 Y}{\partial \phi^2}\right)}{Y}}_{\text{Solutions are spherical harmonics } Y^\ell_m(\theta,\phi)!} = \lambda$$

This only works if $V(r,\theta,\phi) \to V(r)$! Conserved eigenvalue λ related to angular momentum

Radial Equation

$$\frac{\partial}{\partial r}\left(r^2 \frac{\partial R}{\partial r}\right) - \frac{2mr^2}{\hbar^2}(V(r)-E)R = \lambda R$$

define $u = rR(r) \rightarrow R = \frac{u}{r}$, $\quad \frac{\partial R}{\partial r} = \frac{\partial}{\partial r}\frac{u}{r} = \frac{r\frac{\partial u}{\partial r} - u}{r^2}$

Then $\frac{\partial}{\partial r}\left(r^2 \frac{\partial R}{\partial r}\right) = \frac{\partial}{\partial r}\left(r \frac{\partial u}{\partial r} - u\right) = r\frac{\partial^2 u}{\partial r^2} + \cancel{\frac{\partial u}{\partial r}} - \cancel{\frac{\partial u}{\partial r}}$

$$r \frac{\partial^2 u}{\partial r^2} - \frac{2mr}{\hbar^2}(V(r)-E)u = \lambda \frac{u}{r}$$

$$-\frac{\hbar^2}{2m}\frac{\partial^2 u}{\partial r^2} + (V(r)-E)u = -\frac{\hbar^2}{2m}\frac{\lambda u}{r^2}$$

$$-\frac{\hbar^2}{2m}\frac{\partial^2 u}{\partial r^2} + \left(V(r) + \frac{\hbar^2 \lambda}{2mr^2}\right)u = Eu$$

↖ centrifugal potential, classically $\frac{L^2}{2mr^2}$
so $L^2 \rightarrow \hbar^2 \ell(\ell+1)$

3D infinite spherical well

$$V(r) = \begin{cases} 0 & r < a \\ \infty & r \geq a \end{cases}$$

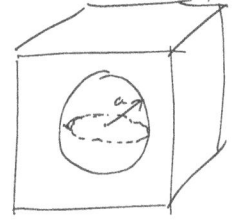

Solve: $\quad -\frac{\hbar^2}{2m}\frac{\partial^2 u}{\partial r^2} + \left(V(r) + \frac{\hbar^2 \ell(\ell+1)}{2mr^2}\right)u = Eu$, $\quad R(r) = \frac{u(r)}{r}$

$r < a$: $\quad \frac{\partial^2 u}{\partial r^2} = \left(\frac{\ell(\ell+1)}{r^2} - \frac{2mE}{\hbar^2}\right)u$, $\quad \ell = 0, 1, 2, \ldots$

For simplest case, $\ell = 0$:

$\frac{\partial^2 u}{\partial r^2} = -\frac{2mE}{\hbar^2}u = -k^2 u \longrightarrow$ $\quad u(r) = A\sin kr + B\cos kr$
$\quad R(r) = \frac{u(r)}{r} = A\frac{\sin kr}{r} + B\frac{\cos kr}{r}$

Boundary conditions:

$R(r=0)$ must be finite to normalize Ψ: $B = 0$.

$R(r=a) = 0 \longrightarrow A\frac{\sin ka}{a} = 0 \longrightarrow ka = n\pi$ so $k_n = \frac{n\pi}{a}$, $n = 1, 2, 3, \ldots$

$E_{n00} = \frac{\hbar^2 n^2 \pi^2}{2ma^2}$, $\Psi_{n00}(r,\theta,\phi) = R_{n0}(r) Y_{00}(\theta,\phi) = A\frac{\sin k_n r}{r} Y_{00} = A'\frac{\sin k_n r}{r}$

Normalization

$$\iiint \psi^* \psi \, d^3r = \int_0^{2\pi}\int_0^{\pi}\int_0^{\infty} \psi^*(r,\theta,\phi)\psi(r,\theta,\phi) r^2 \sin\theta \, dr \, d\theta \, d\phi = 1$$

By convention we normalize angular and radial parts separately:

$$\int_0^{2\pi}\int_0^{\pi} Y_{\ell m}^*(\theta,\phi) Y_{\ell m}(\theta,\phi) \sin\theta \, d\theta \, d\phi = 1$$

$$\int_0^{\infty} R_{n\ell}^*(r) R_{n\ell}(r) r^2 \, dr = 1$$

TABLE 2.1 SPHERICAL HARMONICS

$l = 0$

$Y_{00} = \frac{1}{\sqrt{4\pi}}$

$l = 1$

$Y_{1,\pm 1} = \mp \sqrt{\frac{3}{8\pi}} \sin\theta \, e^{\pm i\phi}$

$Y_{10} = \sqrt{\frac{3}{4\pi}} \cos\theta$

$l = 2$

$Y_{2,\pm 2} = \sqrt{\frac{15}{32\pi}} \sin^2\theta \, e^{\pm 2i\phi}$

$Y_{2,\pm 1} = \mp \sqrt{\frac{15}{8\pi}} \sin\theta \cos\theta \, e^{\pm i\phi}$

$Y_{20} = \sqrt{\frac{5}{16\pi}} (2\cos^2\theta - \sin^2\theta)$

$l = 3$

$Y_{3,\pm 3} = \mp \sqrt{\frac{35}{64\pi}} \sin^3\theta \, e^{\pm 3i\phi}$

$Y_{3,\pm 2} = \sqrt{\frac{105}{32\pi}} \sin^2\theta \cos\theta \, e^{\pm 2i\phi}$

$Y_{3,\pm 1} = \mp \sqrt{\frac{21}{64\pi}} (4\cos^2\theta \sin\theta - \sin^3\theta) \, e^{\pm i\phi}$

$Y_{30} = \sqrt{\frac{7}{16\pi}} (2\cos^3\theta - 3\cos\theta \sin^2\theta)$

l≠0 solutions

$$R_{n\ell}(r) = A_{n\ell} \, j_\ell(k_{n\ell} r)$$

Bessel function of order ℓ

n^{th} zero of the ℓth Bessel function $\beta_{n\ell}$ satisfies B.C.'s: $k_{n\ell} = \frac{\beta_{n\ell}}{a}$

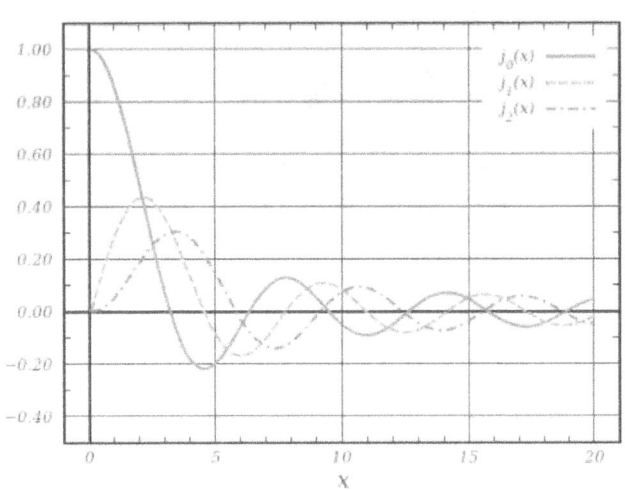

83

Numerical results

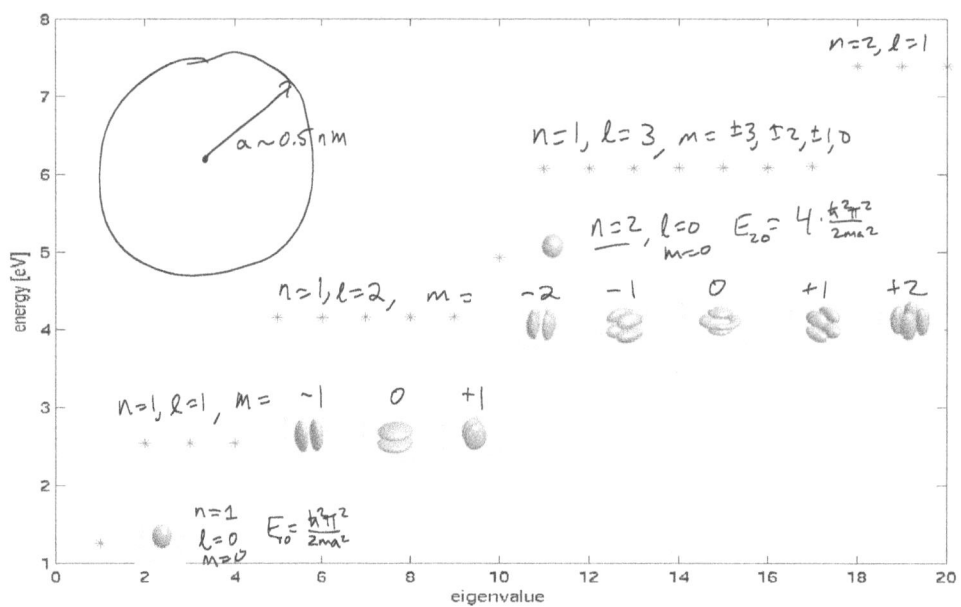

Atomic spectrum of hydrogen

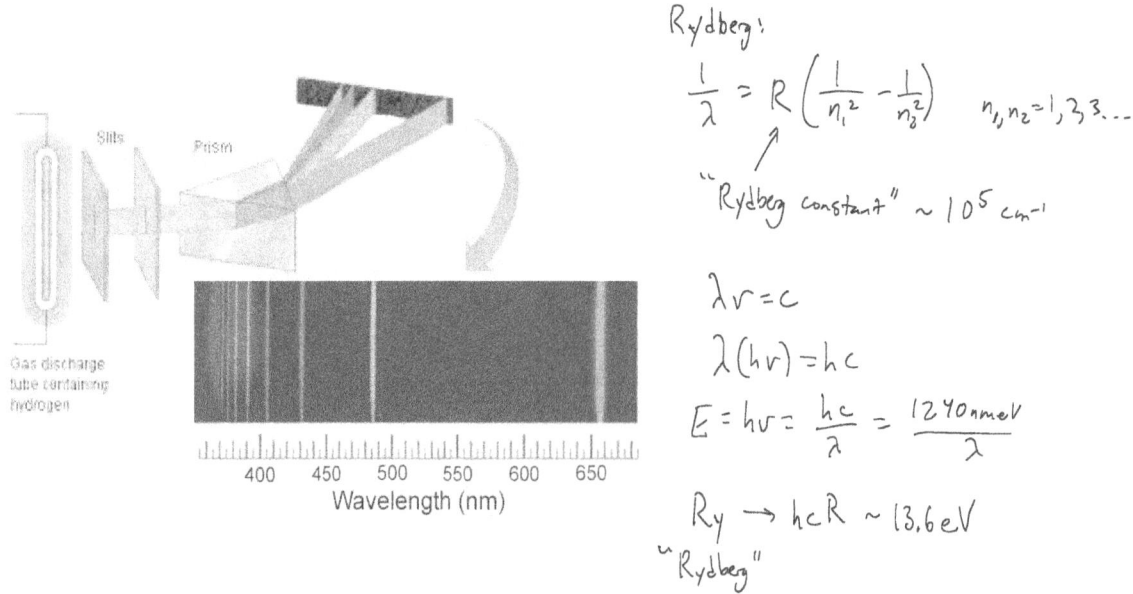

Rydberg:
$$\frac{1}{\lambda} = R\left(\frac{1}{n_1^2} - \frac{1}{n_2^2}\right) \quad n_1, n_2 = 1, 2, 3, \ldots$$

"Rydberg constant" $\sim 10^5 \, cm^{-1}$

$\lambda \nu = c$

$\lambda (h\nu) = hc$

$E = h\nu = \frac{hc}{\lambda} = \frac{1240 \, nm\,eV}{\lambda}$

$R_y \to hcR \sim 13.6 \, eV$

"Rydberg"

Classical Rutherford model

$$E_{tot} = E_{kinetic} + E_{potential} = \tfrac{1}{2}mv^2 + \left(-\frac{e^2}{4\pi\epsilon_0 r}\right) = \tfrac{1}{2}mv^2 - mv^2 = -\tfrac{1}{2}mv^2$$

$$F_{centripetal} = F_{electrostatic} \rightarrow \frac{mv^2}{r} = \frac{e^2}{4\pi\epsilon_0 r^2} \rightarrow r = \frac{e^2}{4\pi\epsilon_0 mv^2}$$

acceleration: $\frac{v^2}{r}$ $\rightarrow$ EM radiation and finite lifetime of atom!

"Semiclassical" Bohr atom

deBroglie: $\lambda = \frac{h}{p}$

Bohr: Circumference of orbit $= 2\pi r = n\lambda = n\frac{h}{mv}$

$\rightarrow mvr = L = n\frac{h}{2\pi} = n\hbar$

$$= mv \frac{e^2}{4\pi\epsilon_0 mv^2} = \frac{e^2}{4\pi\epsilon_0 v} = n\hbar \rightarrow v = \frac{e^2}{4\pi\epsilon_0 n\hbar}$$

$$E_{tot} = -\tfrac{1}{2}mv^2 = -\tfrac{1}{2}m\left(\frac{e^2}{4\pi\epsilon_0 n\hbar}\right)^2 \simeq -\frac{13.6\,eV}{n^2}$$

$$r = \frac{e^2}{4\pi\epsilon_0 mv^2} = \frac{4\pi\epsilon_0 \hbar^2 n^2}{me^2} = a_0 n^2$$

↑ "Bohr radius" $\sim 0.5\,\text{Å}$ (5×10^{-9} cm)

Bohr's hydrogen spectrum

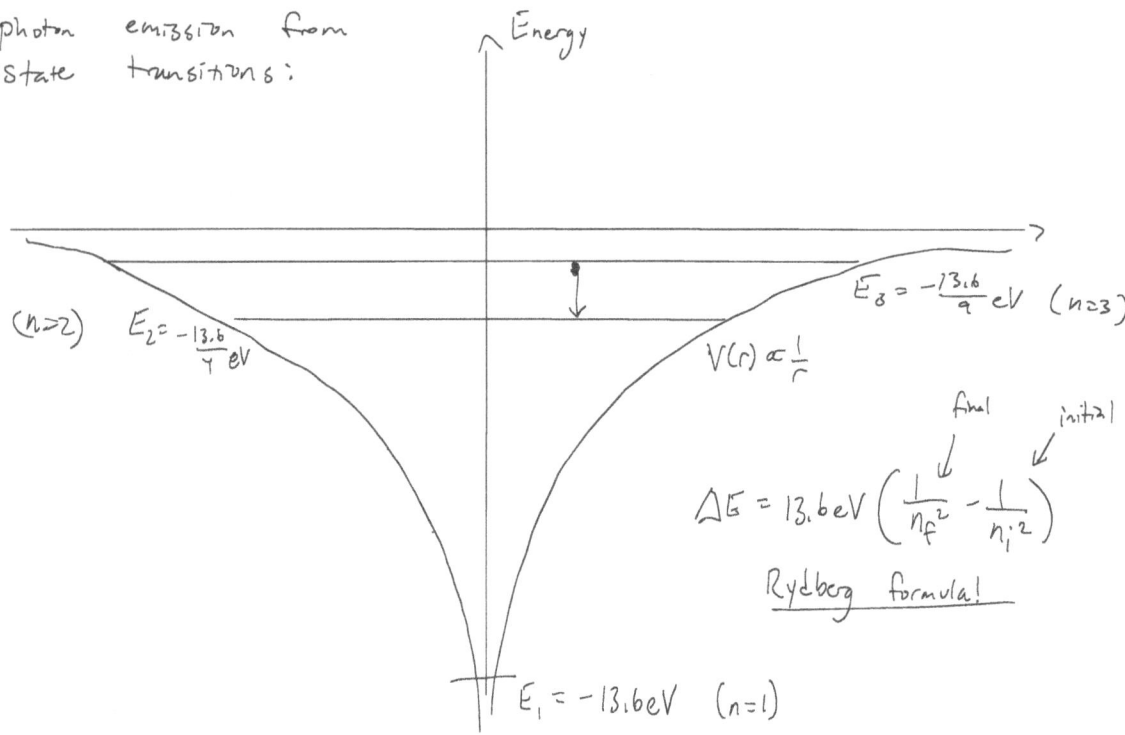

photon emission from state transitions:

$\Delta E = 13.6 \, eV \left(\frac{1}{n_f^2} - \frac{1}{n_i^2} \right)$

(final, initial)

Rydberg formula!

Problems with Bohr's model

1. Orbit w/ definite radius violates uncertainty principle!

2. Angular momentum of ground state is nonzero!

3. Fails to explain:
    a. spectra of larger atoms w/ >1 electron
    b. multiplets (zero-field splitting of spectral lines)
    c. magnetic field-induced spectral line splitting (Zeeman effect)

QM of hydrogen atom

$$V(r) = -\frac{e^2}{4\pi\epsilon_0 r} \quad \text{(spherically symmetric)} \longrightarrow \psi_{n\ell m}(r,\theta,\phi) = R_{n\ell}(r) Y_{\ell m}(\theta,\phi)$$

Radial equation:
$$-\frac{\hbar^2}{2m}\frac{d^2 u}{dr^2} + \left[-\frac{e^2}{4\pi\epsilon_0 r} + \frac{\hbar^2 \ell(\ell+1)}{2m r^2}\right] u = E u, \quad u(r) \equiv r R(r)$$

$$\kappa^2 \equiv -\frac{2mE}{\hbar^2}$$

$$\frac{1}{\kappa^2}\frac{d^2 u}{dr^2} + \left[\frac{e^2}{4\pi\epsilon_0 r \frac{\hbar^2 \kappa^2}{2m}} - \frac{\ell(\ell+1)}{\kappa^2 r^2} - 1\right] u = 0$$

$$\rho \equiv \kappa r \longrightarrow d\rho = \kappa\, dr$$

$$\frac{d^2 u}{d\rho^2} + \left[\left(\frac{2m e^2}{4\pi\epsilon_0 \hbar^2 \kappa}\right)\frac{1}{\rho} - \frac{\ell(\ell+1)}{\rho^2} - 1\right] u = 0$$

$$\frac{d^2 u}{d\rho^2} + \left[\frac{\rho_0}{\rho} - \frac{\ell(\ell+1)}{\rho^2} - 1\right] u = 0$$

Asymptotic behavior of u(ρ)

$\rho \to \infty:$ $\quad \sim \frac{d^2 u}{d\rho^2} - u = 0 \longrightarrow u(\rho) = A e^{-\rho} + \cancel{B e^{+\rho}}\; \to 0$ for normalizable wavefunction
Boundary condition $e^{-\rho} \to \infty$

$\rho \to 0:$ $\quad \sim \frac{d^2 u}{d\rho^2} - \frac{\ell(\ell+1)}{\rho^2} u = 0 \longrightarrow u(\rho) = C \rho^{\ell+1} + \cancel{D \rho^{-\ell}}\; \to 0$
Boundary condition $e^{-\rho} = 0$

[Graph showing $u(\rho)$ vs ρ, with curve $\propto e^{\rho}$]

Ansatz: $\quad u(\rho) = \rho^{\ell+1} e^{-\rho} v(\rho)$

Second derivative of u(ρ)

$$u(\rho) = \rho^{\ell+1}\left(e^{-\rho} v(\rho)\right) = f \cdot g$$

$$(f \cdot g)' = f'g + g'f$$
$$(f \cdot g)'' = f''g + 2g'f' + g''f$$

$$f' = (\ell+1)\rho^\ell, \quad f'' = (\ell+1)\ell \rho^{\ell-1}$$

$$g' = v'e^{-\rho} - e^{-\rho}v, \quad g'' = e^{-\rho}v - 2e^{-\rho}v' + e^{-\rho}v''$$

$$u'' = \ell(\ell+1)\rho^{\ell-1} e^{-\rho}v + 2(\ell+1)\rho^\ell (v'e^{-\rho} - e^{-\rho}v) + \rho^{\ell+1}\left(e^{-\rho}v - 2e^{-\rho}v' + e^{-\rho}v''\right)$$

$$= \rho^\ell e^{-\rho}\left[\rho v'' + (2(\ell+1) - 2\rho)v' + \left(\frac{\ell(\ell+1)}{\rho} - 2(\ell+1) + \rho\right)v\right]$$

Equation for V(ρ)

$$u'' + \left[\frac{\rho_0}{\rho} - \frac{\ell(\ell+1)}{\rho^2} - 1\right]u = 0, \quad u = \rho^{\ell+1}e^{-\rho}v = \rho^\ell e^{-\rho}(\rho v)$$

$$\rho^\ell e^{-\rho}\left[\rho v'' + (2(\ell+1) - 2\rho)v' + \left(\frac{\ell(\ell+1)}{\rho} - 2(\ell+1) + \rho\right)v \right.$$
$$\left. + \left(\rho_0 - \frac{\ell(\ell+1)}{\rho} - \rho\right)v\right] = 0$$

$$\rho v'' + 2(\ell+1-\rho)v' + (\rho_0 - 2(\ell+1))v = 0$$

Due to non-constant coefs, must solve via "brute-force"!

Brute force solution

$$V(\rho) = \sum_{j=0}^{\infty} c_j \rho^j$$

$$\frac{dV(\rho)}{d\rho} = \sum_{j=0}^{\infty} j c_j \rho^{j-1} = \sum_{j=0}^{\infty} (j+1) c_{j+1} \rho^j$$

$$\frac{d^2V(\rho)}{d\rho^2} = \sum_{j=0}^{\infty} j(j+1) c_{j+1} \rho^{j-1}$$

$\rho V'' + 2(\ell+1-\rho) V' + (\rho_0 - 2(\ell+1)) V = 0$ now becomes:

$$\sum_{j=0}^{\infty} \left[j(j+1) c_{j+1} + 2(\ell+1)(j+1) c_{j+1} - 2j c_j + (\rho_0 - 2(\ell+1)) c_j \right] \rho^j = 0$$

Only if all coefs are zero:

$$j(j+1) c_{j+1} + 2(\ell+1)(j+1) c_{j+1} - 2j c_j + (\rho_0 - 2(\ell+1)) c_j = 0$$

Recursion relation

$$c_{j+1} = \frac{2(j+\ell+1) - \rho_0}{(j+1)(j+2\ell+2)} c_j$$

For large j, $c_{j+1} \sim \frac{2}{j+1} c_j \longrightarrow c_j = \frac{2^j}{j!} c_0$

Then, $V(\rho) = \sum_{j=0}^{\infty} c_j \rho^j = \sum_{j=0}^{\infty} c_0 \frac{(2\rho)^j}{j!} = c_0 e^{2\rho}$

So $u(\rho) = \rho^{\ell+1} e^{-\rho} V(\rho) = c_0 \rho^{\ell+1} e^{\rho}$ Not normalizable!

So series must terminate with $c_{j_{max}+1} = 0 \longrightarrow V(\rho)$ is a polynomial!

Principle quantum number

$$2(j_{max}+l+1) - \rho_0 = 0 \text{ for some } j_{max}$$

Since integers $j \geq 0$ and $l \geq 0$, values of $j_{max} + l + 1 = 1, 2, 3, \ldots \equiv n$

"principle quantum number"

This constrains possible values of $l = n - j_{max} - 1 \geq 0$

for $n=1$, $j_{max} = 0 \rightarrow l = 0$

for $n=2$, $j_{max} = 0, 1 \rightarrow l = 1, 0$

for $n=3$, $j_{max} = 0, 1, 2 \rightarrow l = 2, 1, 0$

$\vdots$

$\rightarrow l = n-1, n-2, \ldots, 1, 0$

Energy

$$2n - \rho_0 = 0 \longrightarrow 2n - \left(\frac{2me^2}{4\pi\epsilon_0 \hbar^2 K}\right) = 0 \longrightarrow K = \frac{me^2}{4\pi\epsilon_0 \hbar^2 n}$$

So

$$E = -\frac{\hbar^2 K^2}{2m} = -\frac{\hbar^2}{2m}\left(\frac{me^2}{4\pi\epsilon_0 \hbar^2 n}\right)^2 = -\frac{m}{2\hbar^2}\left(\frac{e^2}{4\pi\epsilon_0}\right)^2 \frac{1}{n^2} \quad n = 1, 2, 3\ldots$$

(Bohr's formula!)

Unlike Bohr's model:
1. orbital angular momentum $l = 0$ for ground state
2. States are n^2 degenerate

Energy level diagram:
- $n=3$, $l=0$, $m=0$; $l=1$, $m=-1,0,1$; $l=2$, $m=-2,-1,0,1,2$
- $n=2$, $l=0$ or $l=1$; $m=0$; $m=-1,0,1$
- $n=1$, $l=0$, $m=0$

Wavefunctions

$$\Psi_{n\ell m}(r,\theta,\phi) = R_{n\ell}(r)\, Y_{\ell m}(\theta,\phi)$$

$$R_{n\ell}(r) = \frac{u_{n\ell}(r)}{r} = \frac{\rho^{\ell+1} e^{-\rho} v_{n\ell}(\rho)}{r}, \quad \rho = Kr = \frac{me^2 r}{4\pi\varepsilon_0 \hbar^2 n} = \frac{r}{a_0 n}$$

$$a_0 = \frac{4\pi\varepsilon_0 \hbar^2}{me^2} \quad \text{Bohr radius}$$

$v_{n\ell}(\rho)$ is determined by sum of polynomial terms given by recursion

Matter is a wave!

- Schrödinger wave eqn: $\mathcal{H}\Psi(x,t) = -\frac{\hbar^2}{2m}\nabla^2\Psi(x,t) + V(x,t)\Psi(x,t) = i\hbar\frac{\partial}{\partial t}\Psi(x,t)$

  is a statement about total energy $\frac{p^2}{2m} + V = E$, obtained by "canonical substitution" of differential operators $p \to \frac{\hbar}{i}\nabla$, $E \to i\hbar\frac{\partial}{\partial t}$

- If $\Psi(x,t) = e^{i(kx-\omega t)}$, then $p\Psi = \frac{\hbar}{i}\nabla e^{i(kx-\omega t)} = \hbar k \Psi$ ("De Broglie", $p = \hbar k$)

- Likewise, $E\Psi = i\hbar\frac{\partial}{\partial t}e^{i(kx-\omega t)} = \hbar\omega\Psi$ ("Einstein", $E = \hbar\omega$)

<u>Fundamental concept</u>: Experimentally measurable "observables" appear as "Hermitian" operators (like the "Hamiltonian" $\mathcal{H}$) acting on a complex-valued "wavefunction" $\Psi(x,t)$ that remains <u>unobservable</u> by direct means!

Interpretation of ψ(x,t)

- Although $\Psi(x,t)$ by itself has no physical meaning, $|\Psi(x,t)|^2$ is interpreted as a probability density ("Born interpretation"). Since total probability = 1, Ψ is "Normalized" by $\int_{-\infty}^{\infty} \Psi^*\Psi\, dx = 1$

- In general, this wavefunction can be any normalized "well-behaved" function that we can expand as a linear combination ("superposition") of orthonormal eigenfunctions ("basis") of a Hermitian operator.

- Measurement of an observable will result in only <u>one</u> eigenvalue of the relevant operator, with probability determined by the squared norm of the complex amplitude of the associated eigenfunction in basis expansion of the wavefunction. Thereafter, the wavefunction has "collapsed" into this state.

- Statistical averages of measurement outcomes are called "expectation values" $\langle O \rangle$

Uncertainty principle

General property of Fourier transform (expansion in basis e^{ikx}): uncertainty ΔK is inversely proportional to Δx: $\Delta K \Delta x \geq 1$

Example: "gaussian"

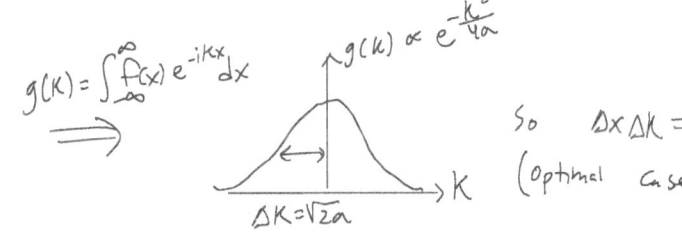

So $\Delta x \Delta K = 1$ (optimal case)

Since $p = \hbar K$ in this basis, and experimental uncertainties are in $|\psi(x)|^2$,
$$\sigma_x \sigma_p \geq \frac{\hbar}{2} \quad \text{(attributed to Heisenberg)}$$
where σ's are uncertainties in expectation values $\sigma_\theta = \sqrt{\langle\theta^2\rangle - \langle\theta\rangle^2}$

More generally, $\sigma_A \sigma_B \geq \left|\frac{1}{2i}[\hat{A},\hat{B}]\right|$, with "commutator" $[\hat{A},\hat{B}] = AB - BA$

ie. There is an uncertainty principle for any pair of "incompatible observables" with non-commuting operators!

Solution of Schrodinger Eqn: simplest cases

$V(x,t) = \text{const}$, boundaries @ $\pm\infty$: "free particle" plane waves $e^{i(kx-\omega t)}$

Whereas the classical wave on string under tension or free EM wave has linear dispersion $\omega(k) = kc$, this quantum wave has a quadratic dispersion $\omega(k) = \frac{\hbar k^2}{2m}$.

But, we still expect familiar wave phenomena! (interference, diffraction, partial reflection, resonance etc.)

$V(x,t) \to V(x)$

PDE is separable into two ODEs (for x and t) with solution
$$\Psi(x,t) = \psi(x) e^{-i\frac{E}{\hbar}t}$$
where $\psi(x)$ is determined by an eigenvalue/eigenfunction equation
$$\mathcal{H}\psi(x) = E\psi(x),$$
Subject to boundary conditions and wavefunction normalization.

Spectrum Quantization

Just like discretization of wave modes in classical closed systems (e.g., wave on string w/ fixed endpoints, EM waves in resonator), "stationary states" of **confined** wavefunctions (eigenfunctions of the Hamiltonian operator) have discrete "quantized" energy eigenvalues.

1D Examples:

- "particle in 1D box" (V const. and finite over finite x):

    $\psi(x)$ sinusoidal / $E \propto n^2$ where n is positive integer "quantum number"

- "harmonic oscillator" ($V \propto x^2$)

    $\psi(x)$ gaussian × Hermite polynomial / $E \propto n$

- Dirac delta ($V = -\alpha \delta(x)$)

    $\psi(x)$: exponential decay / 1 bound state $E \propto -\alpha^2$

Particle probability transport: "scattering"

Conservation of particle probability $|\psi|^2$ leads to probability "current"/"flux" $J = \frac{\hbar}{m} \text{Im}\{\psi^* \nabla \psi\}$. (Similar to classical energy flux in E&M: $\frac{1}{2\omega\mu_0} \text{Im}\{\vec{E}_\perp^* \nabla \vec{E}_\perp\}$)

Example: planewave $\psi_0 e^{ikx}$, $J = \frac{\hbar}{m} \text{Im}\{\psi_0^* e^{-ikx} ik \psi_0 e^{ikx}\} = \frac{\hbar k}{m} \psi_0^* \psi_0 = \underbrace{\frac{\hbar k}{m}}_{\text{velocity}} \cdot \underbrace{\psi_0^* \psi_0}_{\text{density}}$

like classical particle flux!

Generic scattering of planewaves:

$\psi_{inc} = e^{ik_i x} \rightarrow$ {scattering potential $V(x)$} $\psi_{trans} = t e^{ik_f x} \rightarrow$

$\leftarrow \psi_{refl} = r e^{-ik_i x}$

For piecewise-constant 1-d $V(x)$, we apply boundary conditions at each interface (continuity of $\psi(x)$ and $\frac{d\psi}{dx}(x)$) to calculate transmission and reflection coefs

$$T = \frac{J_{trans}}{J_{inc}} = \frac{k_f}{k_i} |t|^2, \quad R = \frac{J_{refl}}{J_{inc}} = |r|^2$$

Phenomena similar to classical waves: interference, resonance, evanescent tunneling...

Approximating continuous space: Finite differences

Using symmetric definition of 1st derivative,

$$\frac{d}{dx}\psi = \lim_{\Delta x \to 0} \frac{\psi(x+\Delta x/2) - \psi(x-\Delta x/2)}{\Delta x} \longrightarrow \frac{d^2}{dx^2}\psi = \lim_{\Delta x \to 0} \frac{\psi(x+\Delta x) - 2\psi(x) + \psi(x-\Delta x)}{\Delta x^2}$$

We can make an approximation: ψ evaluated <u>only</u> at discrete values of x:

$$\cdots \; x_1 \; x_2 \; \cdots \; x_{i-1} \; x_i \; x_{i+1} \; \cdots \; x_{N-1} \; x_N \longrightarrow x$$

$$\frac{d^2}{dx^2}\psi(x_i) \approx \frac{\psi(x_{i-1}) - 2\psi(x_i) + \psi(x_{i+1})}{\Delta x^2} \equiv \frac{\psi_{i-1} - 2\psi_i + \psi_{i+1}}{\Delta x^2}$$

By writing the Schrödinger equation for each value of $i = 1, \ldots, N$, we get the linear system

$$\left[-\frac{\hbar^2}{2m}\frac{d^2}{dx^2} + V(x)\right]\psi(x) \implies \left\{-\frac{\hbar^2}{2m\Delta x^2}\begin{bmatrix} -2 & 1 & 0 & 0 & \\ 1 & -2 & 1 & & \\ 0 & 1 & -2 & 1 & 0 \\ & & & \ddots & \end{bmatrix} + \begin{bmatrix} \ddots & & 0 & & \\ & V_{i-1} & & 0 & \\ 0 & & V_i & & 0 \\ & 0 & & V_{i+1} & \\ & & 0 & & \ddots \end{bmatrix}\right\} \begin{bmatrix} \psi_1 \\ \vdots \\ \psi_{i-1} \\ \psi_i \\ \psi_{i+1} \\ \vdots \\ \psi_N \end{bmatrix} = E \begin{bmatrix} \psi_1 \\ \vdots \\ \psi_{i-1} \\ \psi_i \\ \psi_{i+1} \\ \vdots \\ \psi_N \end{bmatrix}$$

a Hermitian matrix eigenvalue eqn! (That can be solved via Matlab!)

Vector/function spaces, bras & kets

This equivalence between continuous and discrete formulations of QM suggests that we should use a hybrid "Dirac" notation to capture both aspects:

Symbol	Continuous (differential)	Discrete (linear algebra)
$\lvert\psi\rangle$ "ket"	function $\psi(x)$	Column vector
$\langle\psi\rvert$ "bra"	function $\psi^*(x)$	Complex-conj. + transpose (row vector)
$\langle\psi_a\lvert\psi_b\rangle$ inner product	integral $\int_{-\infty}^{\infty} \psi_a^* \psi_b \, dx$	dot product $[\psi_a] \cdot \begin{bmatrix}\psi_b\end{bmatrix}$
$\langle\psi_a\lvert\hat{O}\rvert\psi_b\rangle$ (when $a=b$ "expectation value")	"overlap integral" $\int_{-\infty}^{\infty} \psi_a^* \hat{O} \psi_b \, dx$	matrix element in basis of orthonormal ψ's: $[\psi_a]\begin{bmatrix}\text{matrix}\\\hat{O}\end{bmatrix}\begin{bmatrix}\psi_b\end{bmatrix}$

3D potentials

Generally:
- Three degrees of freedom $\to$ three quantum numbers
  $\to$ "degeneracy": identical energy eigenvalues for eigenfunctions w/ different quantum numbers (both by symmetry and accident)
- Bound states not guaranteed in finite potentials!

For special case of Radial potentials $V(r)$:
- Spherical symmetry $\to$ separable Schrodinger eqn: $\Psi(r,\theta,\phi) = R_{n\ell}(r) Y_\ell^m(\theta,\phi)$ where $Y_\ell^m(\theta,\phi)$ are "spherical harmonics" $\to$ <u>not</u> unique to QM!
- Quantum number $\ell = 0, 1, \ldots$ determines angular momentum and appears in the radial equation for $R(r)$ as a centrifugal potential $\frac{\hbar^2 \ell(\ell+1)}{2mr^2}$
- Important example: One-electron atom, Coulomb potential $V \propto -\frac{1}{r}$: $E \propto -\frac{1}{n^2}$, degeneracy n^2 in ℓ and m, $\ell = 0, \ldots, n-1$ and $m = -\ell, \ldots, \ell$ in integer increments.

Plan of topics for 402

- Atomic Physics
  - Zeeman effect, spin, spin-orbit, hyperfine coupling
  - addition of angular momentum, Clebsch-Gordan coefs.
  - transition selection rules
  - Pauli exclusion and exchange interaction
- Condensed Matter
  - Free electron gas
  - periodic potentials, bandstructure
- Electromagnetism: scalar and vector potentials
- Approximation methods
  - Perturbation Theory { time independent; time dependent $\to$ Fermi golden rule; degenerate
  - WKB
  - Adiabatic approx.
  - Variational principle
- Scattering from 3D central (radial) potentials: Born Approx.

Splitting level degeneracy: Zeeman interaction

Classically, a charged electron with orbital angular momentum forms a closed loop of current and so has a magnetic moment.

$$M = I \cdot \text{Area} = -\frac{ev}{2\pi r}\pi r^2 = -\frac{evr}{2} = -\frac{e(mvr)}{2m} = -\frac{e\hbar}{2m}\frac{|\vec{L}|}{\hbar} = -\mu_B \frac{|\vec{L}|}{\hbar} \quad \text{"Bohr magneton"}$$

(QM: $\sqrt{\ell(\ell+1)}\hbar$)

This magnetic moment interacts with an applied magnetic field and contributes a potential energy

$$E = -\vec{\mu}\cdot\vec{B} = -\mu_z B_z \quad \left(\text{assuming } \vec{B} = B_z \hat{z}, \text{ and } \vec{\mu} = \mu_x \hat{x} + \mu_y \hat{y} + \mu_z \hat{z}\right)$$

So, if states w/ the same principle quantum number n have different $\langle \mu_z \rangle$, their energy eigenvalues will shift differently in a magnetic field and n^2 degeneracy is broken!

Orbital angular momentum in z-direction: L_z

We expect $\langle \mu_z \rangle = -\mu_B \frac{\langle L_z \rangle}{\hbar}$. What is $\langle L_z \rangle$?

Based on our knowledge of <u>linear</u> momentum operators $p_x = \frac{\hbar}{i}\frac{d}{dx}$, $p_y = \frac{\hbar}{i}\frac{d}{dy}$, $p_z = \frac{\hbar}{i}\frac{d}{dz}$, we conclude that the angular momentum z-component operator is given by $\frac{\hbar}{i}$ times the derivative with respect to the conjugate coordinate: $L_z = \frac{\hbar}{i}\frac{d}{d\phi}$. Then, since the ϕ dependence of wavefunctions is given by spherical harmonics,

$$\langle L_z \rangle = \langle Y_{n\ell m} | \frac{\hbar}{i}\frac{\partial}{\partial \phi} | Y_{n\ell m} \rangle$$

$$= \frac{\hbar}{i}\langle e^{-im\phi} | \frac{\partial}{\partial \phi} | e^{im\phi}\rangle = \hbar m \quad \leftarrow \text{"magnetic quantum number"}$$

So Zeeman interaction energy is

$$E = -\vec{\mu}\cdot\vec{B} = -\langle \mu_z \rangle B_z = \mu_B \frac{\langle L_z \rangle}{\hbar} B_z = \mu_B m B_z$$

"Normal" Zeeman effect

$\mu_B \sim 5.8 \times 10^{-5}$ eV/T, and maximum terrestrial magnetic fields are ~ 10 T, so Zeeman splitting is small in comparison to Rydberg energy scale $-\frac{13.6 \text{eV}}{n^2}$

[Energy level diagram showing:
- $n=3$: $\ell=0, m=0$; $\ell=1, m=1,0,-1$; $\ell=2, m=2,1,0,-1,-2$
- $n=2$: $\ell=0, m=0$; $\ell=1, m=1,0,-1$
- $n=1$: $\ell=0, m=0$]

- But, we typically don't directly observe the level spectrum... just the energy differences of radiative transitions.
  Are all transitions between arbitrary initial and final states allowed?

Electronic dipole transitions

During transition between initial state $\psi_{n\ell m}$ and final state $\psi_{n'\ell'm'}$, the charged electron is in a superposition

$$\psi \propto \psi_{n\ell m} e^{-i\frac{E_n}{\hbar}t} + \psi_{n'\ell'm'} e^{-i\frac{E_{n'}}{\hbar}t}$$

This state has a dipole moment $-e\langle\vec{r}\rangle$: $\quad \left(\omega \equiv \frac{E_n - E_{n'}}{\hbar}\right)$

$$-e\langle\psi|\vec{r}|\psi\rangle = -e\int \vec{r}\, \{|\psi_{n\ell m}|^2 + |\psi_{n'\ell'm'}|^2 + \psi^*_{n'\ell'm'}\psi_{n\ell m} e^{-i\omega t} + \psi^*_{n\ell m}\psi_{n'\ell'm'} e^{+i\omega t}\}\, d^3\vec{r}$$

The probability densities are symmetric with respect to spatial inversion, but $\vec{r}$ is antisymmetric, so the first two terms are identically zero:

$$= -e\int \vec{r}\, \{\psi^*_{n'\ell'm'}\psi_{n\ell m} e^{-i\omega t} + \psi^*_{n\ell m}\psi_{n'\ell'm'} e^{+i\omega t}\}\, d^3\vec{r}$$

Unless this integration over spatial coordinates yields zero identically, the dipole oscillation in time ($e^{\pm i\omega t}$) will radiate electromagnetic waves (photons)

Orbital selection rules

Under what conditions does the azimuthal integral vanish?

Calculate Cartesian components of dipole vector:

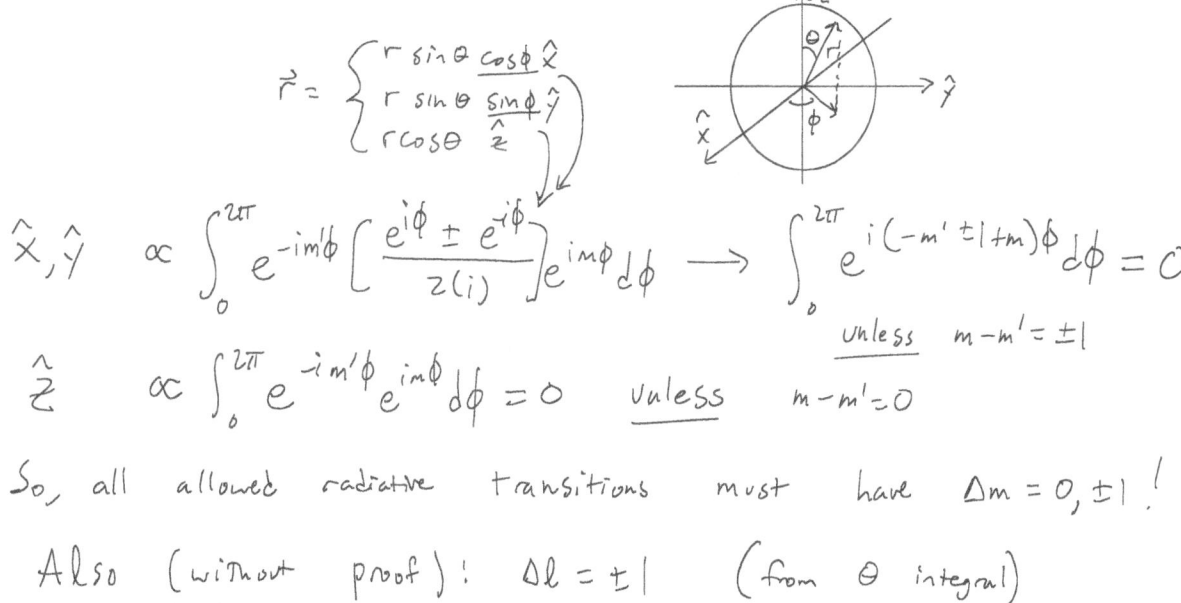

$$\vec{r} = \begin{cases} r\sin\theta \cos\phi\, \hat{x} \\ r\sin\theta \sin\phi\, \hat{y} \\ r\cos\theta\, \hat{z} \end{cases}$$

$$\hat{x},\hat{y} \propto \int_0^{2\pi} e^{-im'\phi}\left[\frac{e^{i\phi} \pm e^{-i\phi}}{2(i)}\right]e^{im\phi}d\phi \longrightarrow \int_0^{2\pi} e^{i(-m' \pm 1 + m)\phi}d\phi = 0$$

$$\underline{\text{unless}} \quad m - m' = \pm 1$$

$$\hat{z} \propto \int_0^{2\pi} e^{-im'\phi} e^{im\phi}d\phi = 0 \quad \underline{\text{unless}} \quad m - m' = 0$$

So, all allowed radiative transitions must have $\Delta m = 0, \pm 1$!

Also (without proof): $\Delta \ell = \pm 1$ (from θ integral)

"Normal" Zeeman effect: dipole-allowed transitions

Only $\Delta \ell = \pm 1$, $\Delta m = \pm 1, 0$ $\longrightarrow$ each line splits into a triplet!

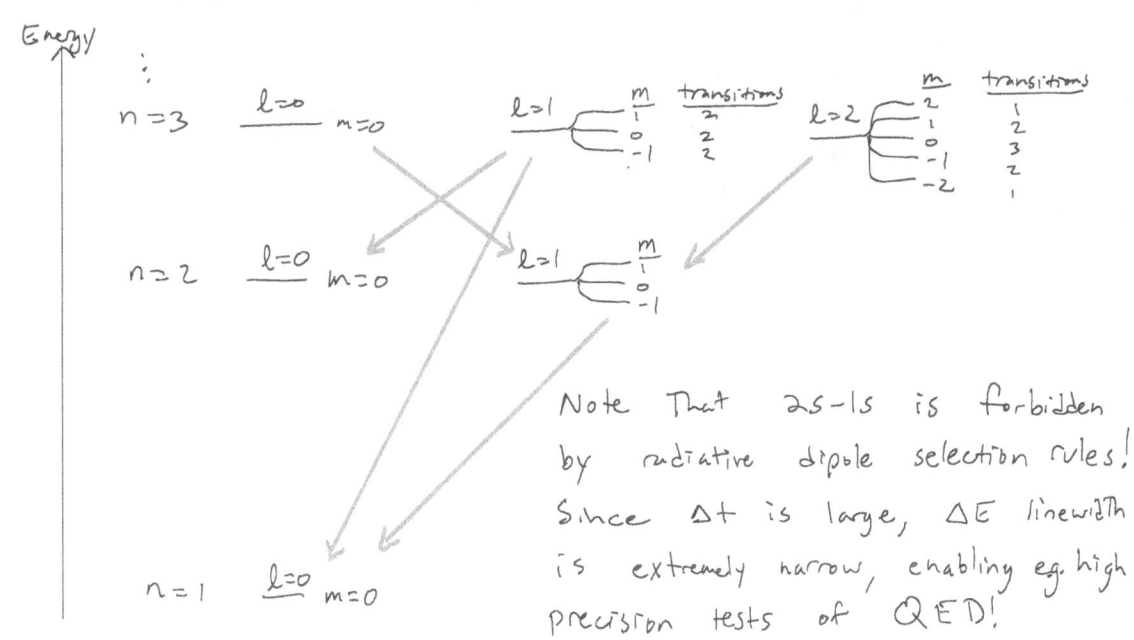

Note that 2s-1s is forbidden by radiative dipole selection rules! Since Δt is large, ΔE linewidth is extremely narrow, enabling e.g. high precision tests of QED!

Polarization

Zeeman-split spectral emission lines are polarized depending on measurement orientation. Since an oscillating dipole only radiates in perpendicular directions:

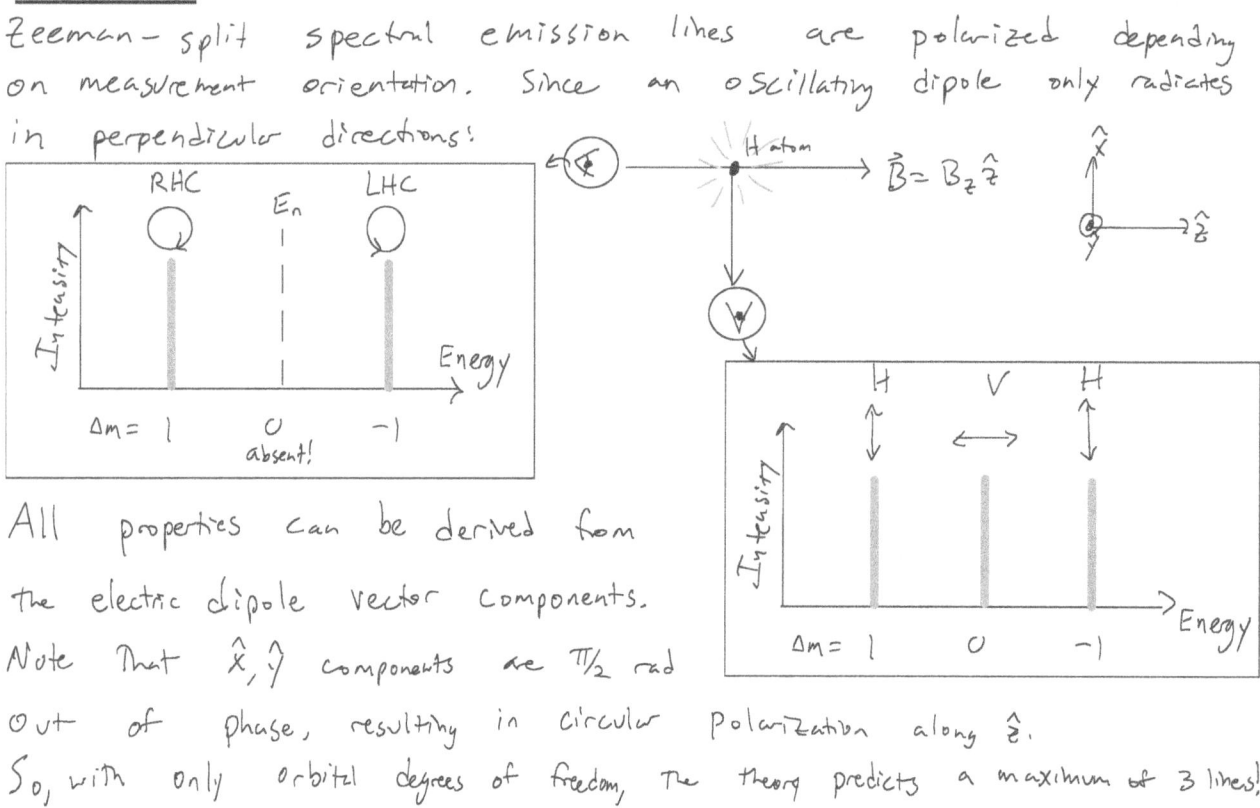

All properties can be derived from the electric dipole vector components. Note that $\hat{x}, \hat{y}$ components are $\pi/2$ rad out of phase, resulting in circular polarization along $\hat{z}$.

So, with only orbital degrees of freedom, the theory predicts a maximum of 3 lines!

Experiment: optical spectroscopy

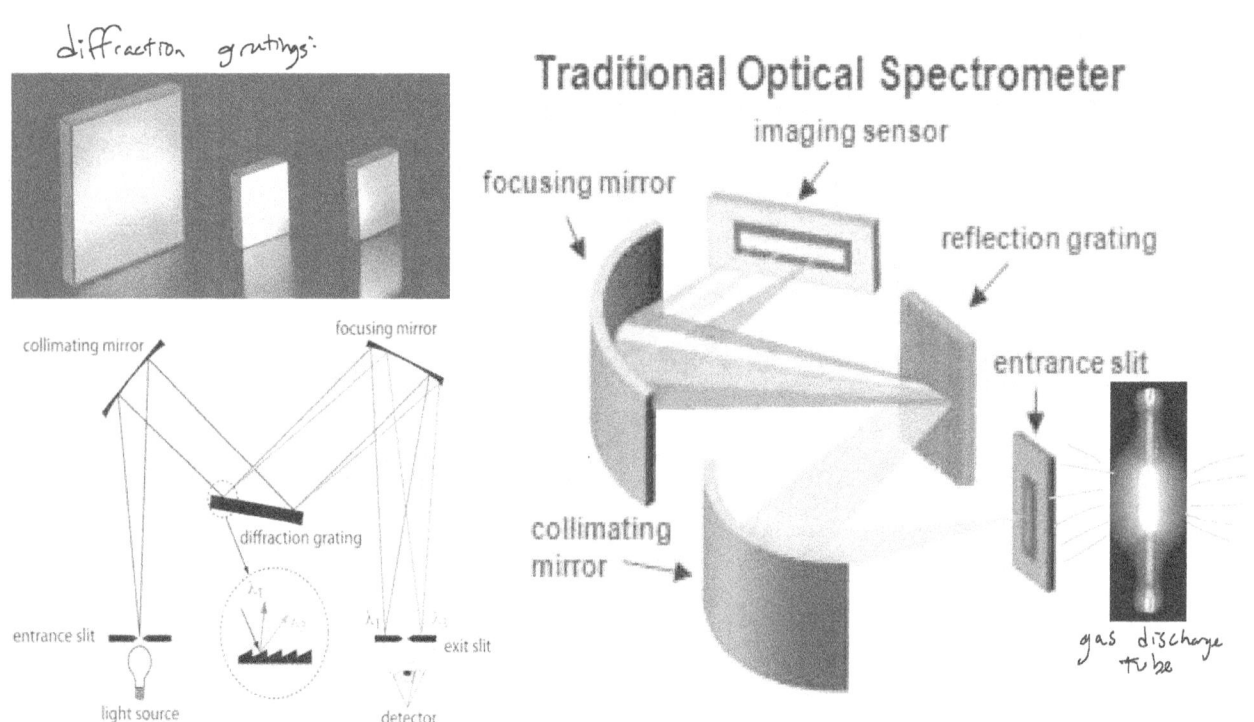

"Anomalous" Zeeman effect

We have predicted that in a magnetic field, each radiative transition line splits into a "triplet" corresponding to $\Delta m = -1, 0, +1$.

However, it is more common to experimentally observe multiplets $\neq 3$! So where is the flaw in our theory?

→ We have ignored the possibility that degrees of freedom other than orbital motion contribute to electron magnetic moment!

But what?

To answer this question, we need to go all the way back to the construction of our wave equation and modify our assumptions implicit in assigning $E = \frac{p^2}{2m} + V$ (!)

Relativistic Quantum Mechanics

Construct a wave equation from the relativistic kinetic energy: (nonrelativistic expression recovered in lowest order!)

$$E = \sqrt{(mc^2)^2 + (pc)^2} = mc^2 \sqrt{1 + \left(\frac{pc}{mc^2}\right)^2} = mc^2\left(1 + \frac{1}{2}\left(\frac{pc}{mc^2}\right)^2 + \ldots\right) \approx mc^2 + \frac{p^2}{2m}$$

Our relativistic wave equation for a free particle is then

$$\sqrt{(mc^2)^2 + (\vec{p}c)^2}\, \Psi = i\hbar \frac{\partial}{\partial t} \Psi$$

But after canonical substitution $\vec{p} \to \frac{\hbar}{i}\nabla$, how to make sense of the square root of an operator?

This problem will disappear if $m^2c^4 + \vec{p}^2c^2$ is a perfect square:

$$\left(\alpha_0 mc^2 + \sum_{j=1}^{3} \alpha_j p_j c\right)^2 = (mc^2)^2 + (\vec{p}c)^2$$

(x,y,z)

This is true only if $\alpha_i^2 = 1$, and $\alpha_i \alpha_j + \alpha_j \alpha_i = \{\alpha_i, \alpha_j\} = 0$, $i \neq j$ ("anti-commutator")

These constraints define a "Clifford algebra", and yield the (free-particle) "Dirac equation":

$$\left[\alpha_0 mc^2 + \sum_{j=1}^{3} \alpha_j p_j c\right] \Psi = i\hbar \frac{\partial}{\partial t} \Psi$$

"Irreducible representation" of "Clifford Algebra"

The anti-commutation relation cannot be satisfied by scalar values — only matrices. The smallest dimension that works is 4x4, and one choice is

$$\alpha_0 = \begin{bmatrix} \hat{I}_2 & \hat{0} \\ \hat{0} & -\hat{I}_2 \end{bmatrix}, \quad \alpha_j = \begin{bmatrix} \hat{0} & \sigma_j \\ \sigma_j & \hat{0} \end{bmatrix},$$

(where $\hat{I}_2$ = 2x2 identity), σ's satisfy commutation rel'n $[\sigma_i, \sigma_j] = 2i\sigma_k$

One choice for σ's are "Pauli matrices": $\sigma_1 = \sigma_x = \begin{bmatrix} 0 & 1 \\ 1 & 0 \end{bmatrix}$, $\sigma_2 = \sigma_y = \begin{bmatrix} 0 & -i \\ i & 0 \end{bmatrix}$, $\sigma_3 = \sigma_z = \begin{bmatrix} 1 & 0 \\ 0 & -1 \end{bmatrix}$

In this (non-unique) basis, the Dirac Hamiltonian is then 4x4:

$$\begin{bmatrix} mc^2 \hat{I}_2 & (\vec{\sigma}\cdot\vec{p})c \\ (\vec{\sigma}\cdot\vec{p})c & -mc^2 \hat{I}_2 \end{bmatrix}, \quad \text{where} \quad \vec{\sigma} = \sigma_x \hat{x} + \sigma_y \hat{y} + \sigma_z \hat{z}.$$

For $p=0$, there are four eigenvalues: **two** are electron-like $(+mc^2)$. The other two $(-mc^2)$ correspond to the "antiparticle" "positron". What discrete degree of freedom do these two wavefunction components correspond to??

Stern-Gerlach experiment (1922)

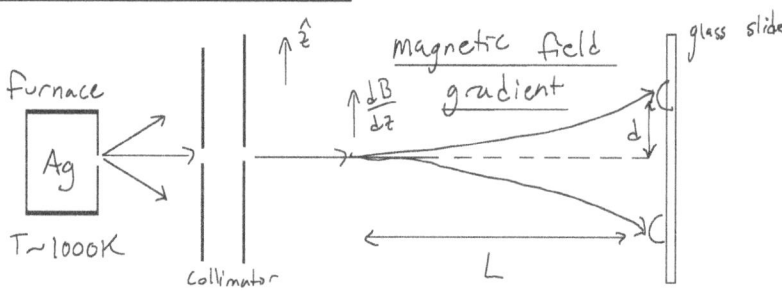

Using $E = -\vec{\mu}\cdot\vec{B}$, neutral atoms with magnetic moment feel a force

$$F = -\frac{dE}{dz} = \vec{\mu}\cdot\frac{d\vec{B}}{dz} = \langle\mu_z\rangle \frac{dB}{dz}.$$

Resulting acceleration causes a displacement over pathlength:

$$d = \frac{1}{2} a t^2 = \frac{1}{2} \frac{F}{m} \left(\frac{L}{v}\right)^2 = \frac{1}{2} \underbrace{\frac{\langle\mu_z\rangle \frac{dB}{dz}}{m}} \frac{L^2}{v^2}$$

The kinetic energy $\sim mv^2$ is determined by thermal energy $k_B T \sim 10^{-1}$ eV. Using $|\mu| = \mu_B$ and realistic parameters,

$$d \approx \frac{1}{6} \frac{\langle\mu_z\rangle \frac{dB}{dz} L^2}{k_B T} = \frac{1}{6} \frac{(5.8 \times 10^{-5} \frac{eV}{T})(10T/cm)(3.5cm)^2}{10^{-1} eV} = 10^{-2} cm = 100 \mu m$$

Experimental results

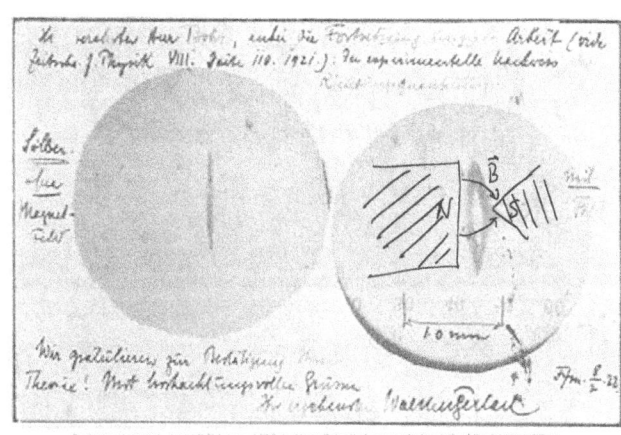

Gerlach's postcard, dated 8 February 1922, to Niels Bohr. It shows a photograph of the beam splitting, with the message, in translation: "Attached [is] the experimental proof of directional quantization. We congratulate [you] on the confirmation of your theory." (Physics Today December 2003).

Since this was done before true QM, S+G didn't know that the total __orbital__ angular momentum of electrons in Ag is ZERO! ($Ag = [Kr]4d^{10}5s^1$) $l=0$
The twofold splitting is therefore due to a magnetic moment associated with __intrinsic__ electron angular momentum, for which there is NO classical analogue! For historical reasons this is called "spin", $\vec{S}$. However, the electron is __NOT__ spinning about any axis.

So, $\langle \mu_z \rangle = -\mu_B \frac{\langle S_z \rangle}{\hbar}$, $S_z = +\frac{\hbar}{2}, -\frac{\hbar}{2} = \hbar m_s$ and $m_s = +\frac{1}{2}, -\frac{1}{2}$. These correspond to the two components in the Dirac wavefunction! By convention these states are often called "spin up" and "spin down" — the vector components are the amplitudes of these two states!

Commutation relations for angular momentum operators

Classically, $\vec{L} = \vec{r} \times \vec{p}$, so $\vec{L} = L_x \hat{x} + L_y \hat{y} + L_z \hat{z}$ where
$L_x = yp_z - zp_y$, $L_y = zp_x - xp_z$, and $L_z = xp_y - yp_x$.
These have a special commutation relation, e.g.
$[L_x, L_y] = [yp_z - zp_y, zp_x - xp_z] = [yp_z, zp_x] - [yp_z, xp_z] - [zp_y, zp_x] + [zp_y, xp_z]$
Since the middle two terms involve momenta along coordinates that don't appear in the commutator, these are zero:

$$= [yp_z, zp_x] + [zp_y, xp_z]$$
$$= yp_z zp_x - zp_x yp_z + zp_y xp_z - xp_z zp_y$$
$$= yp_x(p_z z) - yp_x(zp_z) + xp_y(zp_z) - xp_y(p_z z)$$
$$= -yp_x[z, p_z] + xp_y[z, p_z]$$
$$= [z, p_z](xp_y - yp_x) = i\hbar L_z$$

Likewise, $[L_i, L_j] = i\hbar L_k$, where i, j, k are any cyclic permutation of x, y, z. This applies to __all__ angular momentum operators, i.e. S_x, S_y, S_z or J_x, J_y, J_z!

103

Spin operator and eigenstates

$\vec{S} = S_x \hat{x} + S_y \hat{y} + S_z \hat{z} = \frac{\hbar}{2}(\sigma_x \hat{x} + \sigma_y \hat{y} + \sigma_z \hat{z})$ where σ's are Pauli matrices.

Like orbital angular momentum $\vec{L}$, $\langle S^2 \rangle = \hbar^2 s(s+1)$ and $\langle S_z \rangle = m_s \hbar$.

However, s (unlike ℓ) is __half-integer__: $s = \frac{1}{2}$, $m_s = \pm \frac{1}{2}$.

Spin couples to magnetic field thru its associated magnetic moment,

$\vec{\mu} = -g \frac{\mu_B}{\hbar} \vec{S}$, via the Hamiltonian

$H = -\vec{\mu} \cdot \vec{B} = g \frac{\mu_B}{\hbar} \vec{S} \cdot \vec{B} = \mu_B \sigma_z B_z$ (assuming $\vec{B} = B_z \hat{z}$)

where the "Thomas g-factor" $g = 2$ is due to relativistic "Lorentz boost". Choice of this Pauli basis makes diagonalization trivial:

$H = \begin{bmatrix} +\mu_B B_z & 0 \\ 0 & -\mu_B B_z \end{bmatrix} \implies$
$E = +\mu_B B_z, \quad |\uparrow\rangle = \begin{bmatrix} 1 \\ 0 \end{bmatrix} e^{-i \frac{\mu_B B_z}{\hbar} t}$ "up"
$E = -\mu_B B_z, \quad |\downarrow\rangle = \begin{bmatrix} 0 \\ 1 \end{bmatrix} e^{+i \frac{\mu_B B_z}{\hbar} t}$ "down"

Note that "spin up" has a magnetic moment oriented antiparallel to the field due to opposite polarity of $\vec{\mu}$ and $\vec{S}$.

State evolution

Example: $|\psi(t=0)\rangle$ is an eigenvector of $S_x = \frac{\hbar}{2}\begin{bmatrix} 0 & 1 \\ 1 & 0 \end{bmatrix} \xrightarrow{\text{diagonalize}}$ $+\frac{\hbar}{2}, \frac{1}{\sqrt{2}}\begin{bmatrix} 1 \\ 1 \end{bmatrix} = |+\rangle$
$-\frac{\hbar}{2}, \frac{1}{\sqrt{2}}\begin{bmatrix} 1 \\ -1 \end{bmatrix} = |-\rangle$

What happens when we prepare $|+\rangle$ and apply $\vec{B} = B_z \hat{z}$?

Decompose into new eigenbasis:

$|+\rangle = \frac{1}{\sqrt{2}} \begin{bmatrix} 1 \\ 1 \end{bmatrix} : \frac{1}{\sqrt{2}}\left(\begin{bmatrix} 1 \\ 0 \end{bmatrix} e^{-i \frac{\mu_B B_z}{\hbar} t} + \begin{bmatrix} 0 \\ 1 \end{bmatrix} e^{+i \frac{\mu_B B_z}{\hbar} t}\right)$

What is $\langle S_x(t) \rangle$?

$\langle +|S_x|+\rangle = \frac{1}{\sqrt{2}} \begin{bmatrix} e^{+i \frac{\mu_B B_z}{\hbar} t} & e^{-i \frac{\mu_B B_z}{\hbar} t} \end{bmatrix} \frac{\hbar}{2}\begin{bmatrix} 0 & 1 \\ 1 & 0 \end{bmatrix} \frac{1}{\sqrt{2}}\begin{bmatrix} e^{-i \frac{\mu_B B_z}{\hbar} t} \\ e^{+i \frac{\mu_B B_z}{\hbar} t} \end{bmatrix}$

$= \frac{\hbar}{4}\left(e^{2i \frac{\mu_B B_z}{\hbar} t} + e^{-2i \frac{\mu_B B_z}{\hbar} t}\right) = \frac{\hbar}{2} \cos\left(\frac{2\mu_B B_z}{\hbar} t\right)$: oscillation @ $\omega = \frac{\Delta E}{\hbar}$!

Likewise, $\langle +|S_y|+\rangle = \frac{\hbar}{2} \sin \frac{2 \mu_B B_z}{\hbar} t$; out of phase by 90°!

This rotation of the (fictitious) vector $\langle S_x \rangle \hat{x} + \langle S_y \rangle \hat{y}$ is the same as classical precession of angular momentum about a force causing a torque!

"gyroscope"

Operator form of orbital angular momentum

So we can write spin-1/2 Zeeman Hamiltonian $g\frac{\mu_B}{\hbar}\vec{S}\cdot\vec{B}$ as a 2×2 matrix. What about spin ≠ 1/2 states, ie. $\ell \neq 0$? (the term "spin" applies to all ang. momentum)
Pauli matrices can be generalized (in Pauli basis w/ definite $\langle L_z \rangle$):

$$S_z = \hbar \begin{pmatrix} s & 0 & 0 & 0 & \cdots & 0 \\ 0 & s-1 & 0 & 0 & \cdots & 0 \\ 0 & 0 & s-2 & \cdots & & 0 \\ \vdots & \vdots & \vdots & \ddots & & \vdots \\ 0 & 0 & 0 & \cdots & & -s \end{pmatrix} \quad S_x = \frac{\hbar}{2}\begin{pmatrix} 0 & b_s & 0 & 0 & \cdots & 0 & 0 \\ b_s & 0 & b_{s-1} & 0 & \cdots & 0 & 0 \\ 0 & b_{s-1} & 0 & b_{s-2} & \cdots & 0 & 0 \\ 0 & 0 & b_{s-2} & 0 & \cdots & 0 & 0 \\ \vdots & & & & \ddots & & \vdots \\ 0 & 0 & 0 & 0 & \cdots & 0 & b_{-s+1} \\ 0 & 0 & 0 & 0 & \cdots & b_{-s+1} & 0 \end{pmatrix} \quad S_y = \frac{\hbar}{2i}\begin{pmatrix} 0 & b_s & 0 & 0 & \cdots & 0 & 0 \\ -b_s & 0 & b_{s-1} & 0 & \cdots & 0 & 0 \\ 0 & -b_{s-1} & 0 & b_{s-2} & \cdots & 0 & 0 \\ 0 & 0 & -b_{s-2} & 0 & \cdots & 0 & 0 \\ \vdots & & & & \ddots & & \vdots \\ 0 & 0 & 0 & 0 & \cdots & 0 & b_{-s+1} \\ 0 & 0 & 0 & 0 & \cdots & -b_{-s+1} & 0 \end{pmatrix}$$

(where $b_j = \sqrt{(s+j)(s+1-j)}$).

All have same eigenvalues, and satisfy commutation relation $[L_i, L_j] = i\hbar L_k$.

eg. $\underline{\ell=1}$: $\quad L_x = \frac{\hbar}{\sqrt{2}}\begin{pmatrix} 0 & 1 & 0 \\ 1 & 0 & 1 \\ 0 & 1 & 0 \end{pmatrix} \quad L_y = \frac{\hbar}{\sqrt{2}i}\begin{pmatrix} 0 & 1 & 0 \\ -1 & 0 & 1 \\ 0 & -1 & 0 \end{pmatrix} \quad L_z = \hbar\begin{pmatrix} 1 & 0 & 0 \\ 0 & 0 & 0 \\ 0 & 0 & -1 \end{pmatrix}$

To make prediction about atomic spectral line splitting in B-field, we need to learn how to combine 2-dimensional (intrinsic) spin space with orbital Hilbert space of $3, 5, \ldots$ dimensions (for $\ell = 1, 2, \ldots$). How to do this?

Addition of angular momentum

Simplest possible case: Spin 1/2 and spin 1/2
What is total angular momentum $J^2 = (\vec{S}^{(1)} + \vec{S}^{(2)})^2 = (\vec{S}^{(1)})^2 + (\vec{S}^{(2)})^2 + 2\vec{S}^{(1)}\cdot\vec{S}^{(2)}$?
We need to construct a larger 4-dimensional Hilbert space from two 2-dimensional spaces. Using a basis of spin up and spin down along $\hat{z}$ for each, use the "direct" or "Kronecker" product to construct a larger basis that spans the space:

$$\underbrace{\begin{bmatrix} \uparrow \\ \downarrow \end{bmatrix}}_{S^{(1)}} \otimes \underbrace{\begin{bmatrix} \uparrow \\ \downarrow \end{bmatrix}}_{S^{(2)}} = \begin{bmatrix} \uparrow\begin{bmatrix} \uparrow \\ \downarrow \end{bmatrix}^{S^{(2)}} \\ \downarrow\begin{bmatrix} \uparrow \\ \downarrow \end{bmatrix} \end{bmatrix}^{S^{(1)}} = \begin{bmatrix} \uparrow\uparrow \\ \uparrow\downarrow \\ \downarrow\uparrow \\ \downarrow\downarrow \end{bmatrix}^{S^{(1)}S^{(2)}}$$

Operators must likewise be constructed with $\otimes$:

$S^{(1)}_{x,y,\text{or }z} \longrightarrow S^{(1)}_{x,y,\text{or }z} \otimes \underbrace{I_2}_{2\times 2 \text{ identity}}$ ex.: $S^{(1)}_x \overset{\hbar}{=} \frac{\hbar}{2}\begin{bmatrix} 0 & 1 \\ 1 & 0 \end{bmatrix} \longrightarrow S^{(1)}_x \otimes I_2 = \frac{\hbar}{2}\begin{bmatrix} 0 & 1 \\ 1 & 0 \end{bmatrix} \otimes \begin{bmatrix} 1 & 0 \\ 0 & 1 \end{bmatrix} = \frac{\hbar}{2}\begin{bmatrix} 0 & 0 & 1 & 0 \\ 0 & 0 & 0 & 1 \\ 1 & 0 & 0 & 0 \\ 0 & 1 & 0 & 0 \end{bmatrix}$

$S^{(2)}_{x,y,\text{or }z} \longrightarrow I_2 \otimes S^{(2)}_{x,y,\text{or }z}$ ex.: $S^{(2)}_x \longrightarrow \begin{bmatrix} 1 & 0 \\ 0 & 1 \end{bmatrix} \otimes \frac{\hbar}{2}\begin{bmatrix} 0 & 1 \\ 1 & 0 \end{bmatrix} = \frac{\hbar}{2}\begin{bmatrix} 0 & 1 & 0 & 0 \\ 1 & 0 & 0 & 0 \\ 0 & 0 & 0 & 1 \\ 0 & 0 & 1 & 0 \end{bmatrix}$

And because operators act only on their own subspace,

$S^{(1)}_x S^{(2)}_x = \underbrace{(S^{(1)}_x \otimes I_2)(I_2 \otimes S^{(2)}_x)}_{} = S^{(1)}_x \otimes S^{(2)}_x$

Two spin-1/2 angular momenta: eigenvectors

$$J^2 = (\vec{S}^{(1)})^2 + (\vec{S}^{(2)})^2 + 2\vec{S}^{(1)}\cdot\vec{S}^{(2)}$$

$$= (S_x^{(1)}\otimes I_2)^2 + (S_y^{(1)}\otimes I_2)^2 + (S_z^{(1)}\otimes I_2)^2$$

$$+ (I_2\otimes S_x^{(2)})^2 + (I_2\otimes S_y^{(2)})^2 + (I_2\otimes S_z^{(2)})^2$$

$$+ 2(S_x^{(1)}\otimes S_x^{(2)} + S_y^{(1)}\otimes S_y^{(2)} + S_z^{(1)}\otimes S_z^{(2)})$$

Matlab:
```
S1x=1/2*[0 1;1 0];
S1y=1/2*[0 -i ; i 0];
S1z=1/2*[1 0; 0 -1];
S2x=S1x;S2y=S1y;S2z=S1z;
```
in units of $\hbar$.

$I_N \rightarrow$ eye(N)
$A\otimes B \rightarrow$ kron(A,B)

```
N1=length(S1z); N2=length(S2z);
S1sqrd=kron(S1x, eye(N2))^2+kron(S1y, eye(N2))^2+kron(S1z, eye(N2))^2;
S2sqrd=kron(eye(N1), S2x)^2+kron(eye(N1), S2y)^2+kron(eye(N1), S2z)^2;
S1dotS2=kron(S1x,S2x)+kron(S1y,S2y)+kron(S1z,S2z);

Jsqrd=S1sqrd+S2sqrd+2*S1dotS2;
disp(Jsqrd)
[v,d]=eig(Jsqrd);
disp(v)
disp(diag(d)')
```
v is eigenvectors and d contains eigenvalues

```
>> disp(v)
↑↑    0         0    1.0000    0
↑↓   -0.7071    0.7071    0    0
↓↑    0.7071    0.7071    0    0
↓↓    0         0         0    1.0000

>> disp(diag(d)')
   0    2    2    2
```

$\frac{1}{\sqrt{2}}(|\downarrow\uparrow\rangle - |\uparrow\downarrow\rangle)$, $\frac{1}{\sqrt{2}}(|\uparrow\downarrow\rangle + |\downarrow\uparrow\rangle)$, $|\uparrow\uparrow\rangle$, $|\downarrow\downarrow\rangle$

"singlet"   "triplet"

eigenvalue $0 = 0(0+1)$ $2 = 1(1+1)$

eigenvalues are just like $\langle L^2\rangle/\hbar^2 = \ell(\ell+1)$ .. but with new quantum num. $j = S^{(1)}+S^{(2)}, \ldots, S^{(1)}-S^{(2)}$

```
>> Jz=kron(S1z,eye(N2))+kron(eye(N1),S2z);
>> disp(diag(v'*Jz*v)')
   0    0    1   -1
```

Just like $\langle L_z\rangle/\hbar = m$ but with new $m_j = -j, -j+1, \ldots, j-1, j$

So states w/ definite total ang. momentum are $|j, m_j\rangle = |0,0\rangle, |1,0\rangle, |1,1\rangle, |1,-1\rangle$: $\hat{J}|j\rangle = \hbar |j\rangle$

Integer spin plus spin-1/2

spin 1/2 ⊗ spin 1 → Just replace one set of 2×2 Pauli spin matrices w/ 3×3.

```
S2x=1/sqrt(2)*[0 1 0; 1 0 1; 0 1 0];
S2y=1/sqrt(2)*i*[0 -1 0; 1 0 -1; 0 1 0];
S2z=diag([1 0 -1]);
```

We expect eigenstates of J^2 labeled by $j = 1+\frac{1}{2} = \frac{3}{2}$ or $1-\frac{1}{2} = \frac{1}{2}$ with $j=3/2$: $m_j = -3/2, -1/2, 1/2, 3/2$ / $j=1/2$: $m_j = \pm 1/2$. In spin 1/2 ⊗ spin 1 basis,

eigenvectors of J^2 are:

```
↑  1     0         0    1.0000  √(1/3)   0         0
↑  0    -0.5774    0    0       0       -0.8165    0
↑ -1    0.8165    0     0.5774  0       -0.5774    0
↓  1    0         0.8165 0     -0.5774   0         0
↓  0    -0.5774   0.8165 0      0        0         0
↓ -1    0         0      0      0        0        1.0000
                         √(2/3)
```

examples:
ℓ, m
$|j=1/2, m_j=1/2\rangle = \sqrt{\frac{2}{3}}|1,1\rangle|\downarrow\rangle - \sqrt{\frac{1}{3}}|1,0\rangle|\uparrow\rangle$

$|j=3/2, m_j=-1/2\rangle = \sqrt{\frac{1}{3}}|1,-1\rangle|\uparrow\rangle + \sqrt{\frac{2}{3}}|1,0\rangle|\downarrow\rangle$

```
0.7500  0.7500  3.7500  3.7500  3.7500  3.7500
```

$\underbrace{\frac{3}{4} = \frac{1}{2}(1+\frac{1}{2})}_{j=1/2}$ $\underbrace{\frac{15}{4} = \frac{3}{2}(1+\frac{3}{2})}_{j=3/2}$

diagonal matrix elements of J_z in this basis:
```
0.5000  -0.5000  1.5000  -0.5000  0.5000  -1.5000
```

⇒ fortunately, this calculation has been completed for many different cases of spin addition, with results (i.e. coefficients of product states) tabulated in the so-called "Clebsch-Gordan" tables!

Cheat-sheet: Clebsch-Gordan tables

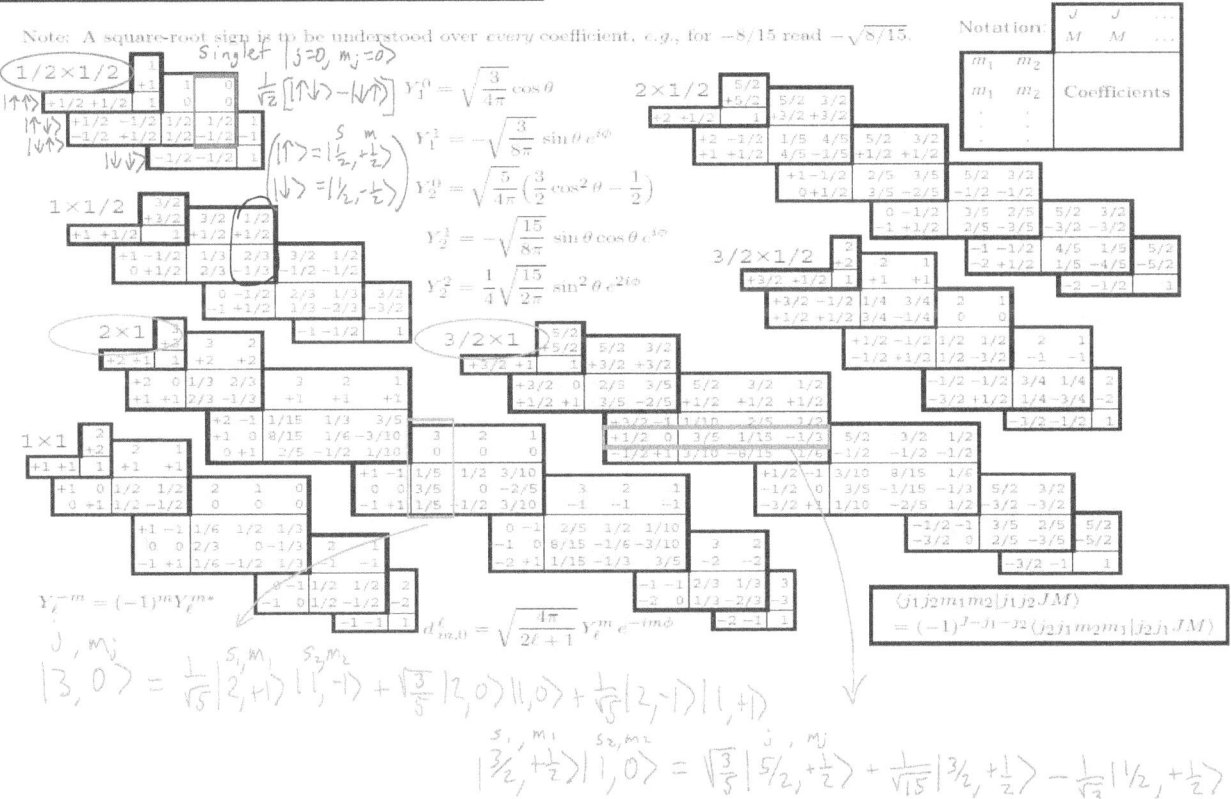

$$|3,0\rangle = \tfrac{1}{\sqrt{5}}|2,+1\rangle|1,-1\rangle + \sqrt{\tfrac{3}{5}}|2,0\rangle|1,0\rangle + \tfrac{1}{\sqrt{5}}|2,-1\rangle|1,+1\rangle$$

$$|\tfrac{3}{2},+\tfrac{1}{2}\rangle|1,0\rangle = \sqrt{\tfrac{3}{5}}|\tfrac{5}{2},+\tfrac{1}{2}\rangle + \tfrac{1}{\sqrt{15}}|\tfrac{3}{2},+\tfrac{1}{2}\rangle - \tfrac{1}{\sqrt{3}}|\tfrac{1}{2},+\tfrac{1}{2}\rangle$$

Spin-orbit coupling

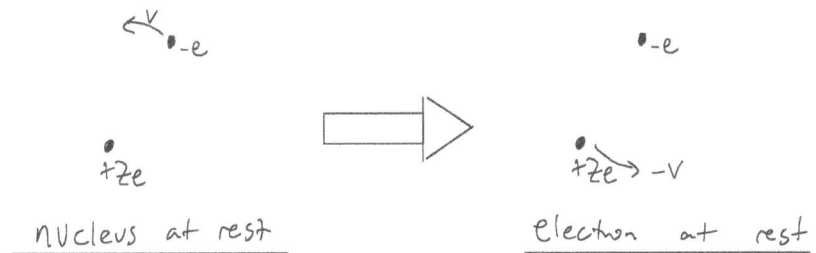

nucleus at rest   →   electron at rest

In the electron's (non-inertial) rest frame, the positively-charged nucleus forms an element of current, creating a local magnetic field via Biot-Savart Law:

$$\vec{B} = \frac{\mu_0}{4\pi}\frac{Ze(-\vec{v}\times\vec{r})}{r^3} = \frac{1}{4\pi\epsilon_0 c^2}\frac{Ze\vec{L}}{mr^3} \quad \left(\text{using } \vec{L} = \vec{r}\times\vec{p} = -m\vec{v}\times\vec{r} \text{ and } \frac{1}{\mu_0\epsilon_0} = c^2\right)$$

this "relativistic" field interacts with the electron's spin:

$$H = -\vec{\mu}\cdot\vec{B} = -\left(-g\frac{\mu_B}{\hbar}\vec{S}\right)\cdot\frac{1}{4\pi\epsilon_0 c^2}\frac{Ze\vec{L}}{mr^3} = \frac{1}{4\pi\epsilon_0}\frac{Ze^2}{2m^2c^2r^3}\vec{S}\cdot\vec{L} \quad \text{"spin-orbit"} \quad \left(\mu_B = \frac{e\hbar}{2m}\right)$$

Estimating Spin-orbit interaction

How big is this spin-orbit interaction energy?

First, look at "<u>gross structure</u>" of the one-electron atom. The ground state energy is

$$E_1 = \frac{-z^2 e^4 m}{2(4\pi\epsilon_0 \hbar)^2} = -\frac{z^2}{2}\left(\frac{e^2}{4\pi\epsilon_0 \hbar c}\right)^2 mc^2 = -\frac{z^2}{2}\alpha^2 E_{rest-mass}$$

where:

$\alpha \equiv \frac{e^2}{4\pi\epsilon_0 \hbar c}$ "fine structure constant" $\sim \frac{1}{137}$ <u>NO</u> units: $\frac{e^2}{\left(\frac{e}{\sqrt{cm}}\right) eV \cdot s \cdot \frac{cm}{s}}$ $\left(\frac{q}{4\pi\epsilon_0 r} = V\right)$

$E_{rest-mass} \equiv mc^2$ "rest mass energy" ~ 511 KeV

"<u>fine structure</u>" (spin-orbit) $E = \frac{1}{4\pi\epsilon_0} \cdot \frac{ze^2}{2m^2c^2}\left\langle \frac{\vec{S}\cdot\vec{L}}{r^3}\right\rangle$

Now, $\langle \frac{1}{r^3}\rangle$ depends on ℓ and n, but can be ballparked in terms of Bohr radius.

$a_0 = \frac{4\pi\epsilon_0 \hbar^2}{ze^2 m}$ and $\vec{S}\cdot\vec{L}$ is on the order of $\hbar^2$.

$E_{spin-orbit} \sim \frac{z^4 e^8 m^3 \hbar^2}{2(4\pi\epsilon_0)^4 m^2 c^2 \hbar^6} = \frac{z^4}{2}\left(\frac{e^2}{4\pi\epsilon_0 \hbar c}\right)^4 mc^2 = \frac{z^4}{2}\alpha^4 E_0$, $\alpha^2 \sim 10^{-4}$ smaller than gross electronic structure.

Zeeman interaction including LS coupling

$H = \frac{\mu_B}{\hbar}\vec{L}\cdot\vec{B} + 2\frac{\mu_B}{\hbar}\vec{S}\cdot\vec{B} + \lambda\vec{L}\cdot\vec{S}$ $\left[\text{For } \vec{B}=B_z\hat{z}, H = \frac{\mu_B}{\hbar}B_z(L_z + 2S_z) + \lambda\vec{L}\cdot\vec{S}\right]$

Using the Kronecker product $\otimes$, we can <u>exactly</u> diagonalize this $2(2\ell+1)\times 2(2\ell+1)$ matrix!

- We see distinct regimes determined by the relative strength of B-dependent Zeeman and B-independent spin-orbit terms with varying B-field magnitude:

For $\underline{\ell=1}$,

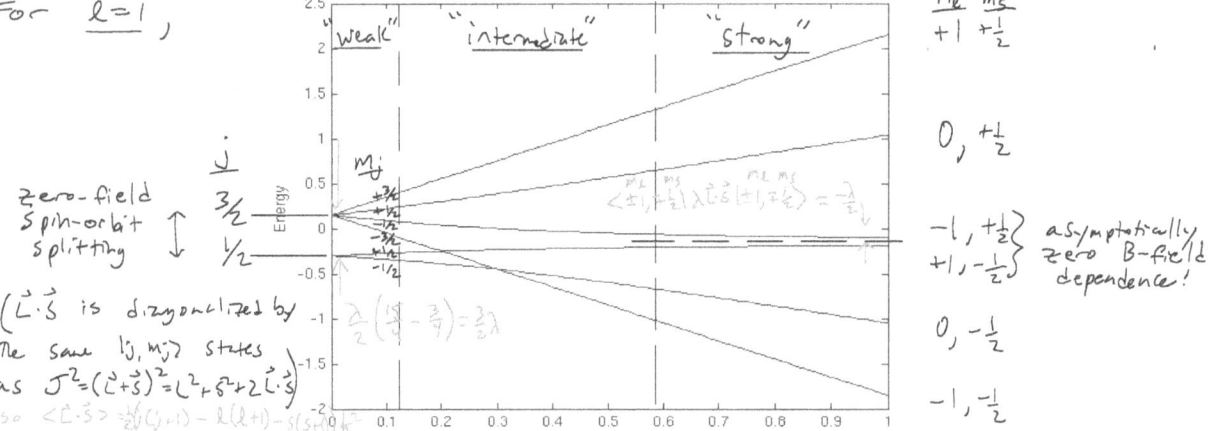

zero-field spin-orbit splitting $\updownarrow$ $\begin{array}{c} j \\ 3/2 \\ 1/2 \end{array}$

($\vec{L}\cdot\vec{S}$ is diagonalized by the same $|j,m_j\rangle$ states as $J^2 = (\vec{L}+\vec{S})^2 = L^2 + S^2 + 2\vec{L}\cdot\vec{S}$ so $\langle \vec{L}\cdot\vec{S}\rangle = \frac{1}{2}[j(j+1) - \ell(\ell+1) - s(s+1)]\hbar^2$)

$\begin{array}{cc} m_\ell & m_s \\ +1 & +\frac{1}{2} \\ 0 & +\frac{1}{2} \\ -1,+\frac{1}{2} \\ +1,-\frac{1}{2} \end{array}$ asymptotically zero B-field dependence!

$0, -\frac{1}{2}$

$-1, -\frac{1}{2}$

- In the <u>strong</u> B regime, Zeeman terms dominate and L_z, S_z are <u>approximately</u> conserved!
- In <u>weak</u> B, $j=3/2, m_j=\pm\frac{1}{2}$ split $2\times$ more than $j=\frac{1}{2}, m_j=\pm 1/2$ states. Why?

Landé g-factor

What are expectation values of the Zeeman terms of the Hamiltonian

$$H = \frac{\mu_B}{\hbar} \vec{L}\cdot\vec{B} + 2\frac{\mu_B}{\hbar}\vec{S}\cdot\vec{B} = \frac{\mu_B}{\hbar}(\vec{L}+2\vec{S})\cdot\vec{B}$$

for the $|j, m_j\rangle$ states that diagonalize $\vec{L}\cdot\vec{S}$ (approximate eigenstates for weak B)? Project components onto $\vec{J}$ unit vector:

$$H = \frac{\mu_B}{\hbar}\frac{(\vec{L}+2\vec{S})\cdot\vec{J}}{|\vec{J}|}\cdot\frac{\vec{B}\cdot\vec{J}}{|\vec{J}|} = \frac{\mu_B}{\hbar}\frac{(\vec{L}+2\vec{S})\cdot(\vec{L}+\vec{S})}{J^2}J_z B_z = \frac{\mu_B}{\hbar}\frac{(L^2+2S^2+3\vec{S}\cdot\vec{L})}{J^2}J_z B_z$$

We know $\langle L^2\rangle = \hbar^2 \ell(\ell+1)$, $\langle S^2\rangle = \hbar^2 s(s+1)$, $\langle J^2\rangle = \hbar^2 j(j+1)$, $\langle J_z\rangle = \hbar m_j$

and $\vec{S}\cdot\vec{L}$ determined by $J^2 = (\vec{L}+\vec{S})^2 = L^2+S^2+2\vec{S}\cdot\vec{L} \longrightarrow \vec{S}\cdot\vec{L} = \frac{J^2-L^2-S^2}{2}$

Then, $E = \langle H\rangle = \frac{\mu_B}{\hbar}\left(\frac{\langle L^2\rangle + 2\langle S^2\rangle + \frac{3}{2}(\langle J^2\rangle - \langle L^2\rangle - \langle S^2\rangle)}{\langle J^2\rangle}\right)\hbar m_j B_z$

$$= \mu_B \frac{\frac{3}{2}\langle J^2\rangle - \frac{1}{2}\langle L^2\rangle + \frac{1}{2}\langle S^2\rangle}{\langle J^2\rangle} m_j B_z = \left(1 + \frac{\langle J^2\rangle - \langle L^2\rangle + \langle S^2\rangle}{2\langle J^2\rangle}\right)\mu_B m_j B_z$$

$$= g_L \mu_B m_j B_z$$

where the "Landé g-factor" $\quad g_L = 1 + \frac{j(j+1) - \ell(\ell+1) + s(s+1)}{2j(j+1)}$

Landé g-factor examples

$\underline{\ell=0}$ $(j=s=\frac{1}{2})$: $\quad g_L = 1 + \frac{\frac{1}{2}(\frac{3}{2}) - 0(0+1) + \frac{1}{2}(\frac{3}{2})}{2\cdot\frac{1}{2}(\frac{3}{2})} = 2$ ✓

$\underline{s=0}$ $(j=\ell)$: $\quad g_L = 1 + \frac{\ell(\ell+1) - \ell(\ell+1) + 0(1+0)}{2\ell(\ell+1)} = 1$ ✓

$\underline{\ell=1}$ $(j=\frac{1}{2}$ or $\frac{3}{2})$: $g_L = \begin{cases} (j=\frac{1}{2}) \;\; 1 + \frac{\frac{1}{2}(\frac{3}{2}) - 1(2) + \frac{1}{2}(\frac{3}{2})}{2\cdot\frac{1}{2}(\frac{3}{2})} = 1 + \frac{\frac{3}{4} - \frac{8}{4} + \frac{3}{4}}{6/4} = \frac{2}{3} \\ \text{OR} \\ (j=\frac{3}{2}) \;\; 1 + \frac{\frac{3}{2}(\frac{5}{2}) - 1(2) + \frac{1}{2}(\frac{3}{2})}{2\cdot\frac{3}{2}(\frac{5}{2})} = 1 + \frac{\frac{15}{4} - \frac{8}{4} + \frac{3}{4}}{30/4} = \frac{4}{3} \end{cases}$

Numerically,

```
Sx=1/2*[0 1;1 0];
Sy=1/2*[0 -i ; i 0];
Sz=1/2*[1 0; 0 -1];
Lx=1/sqrt(2)*[0 1 0; 1 0 1; 0 1 0];
Ly=1/sqrt(2)*[0 -i 0; i 0 -i; 0 i 0];
Lz=[1 0 0; 0 0 0 ; 0 0 -1];
Ssqrd=kron(eye(3), Sx)^2+kron(eye(3), Sy)^2+kron(eye(3), Sz)^2;
Lsqrd=kron(Lx, eye(2))^2+kron(Ly, eye(2))^2+kron(Lz, eye(2))^2;
SdotL=kron(Lx,Sx)+kron(Ly,Sy)+kron(Lz,Sz);
Jsqrd=Ssqrd+Lsqrd+2*SdotL; [v,d]=eig(Jsqrd);
gL=(Lsqrd+2*Ssqrd+3*SdotL)/Jsqrd;
disp([diag(v'*Jsqrd*v) diag(v'*gL*v)])
```

j(j+1)	g_L
0.7500	0.6667
0.7500	0.6667
3.7500	1.3333
3.7500	1.3333
3.7500	1.3333
3.7500	1.3333

$j=1/2, g_L=2/3$ ✓

$j=3/2, g_L=4/3$ ✓

Anomalous Zeeman effect explained

Selection rules $\Delta j = \pm 1, 0$, $\Delta m_j = 0, \pm 1$ limit possible transitions.
For sodium "doublet" (split by zero-field spin orbit):

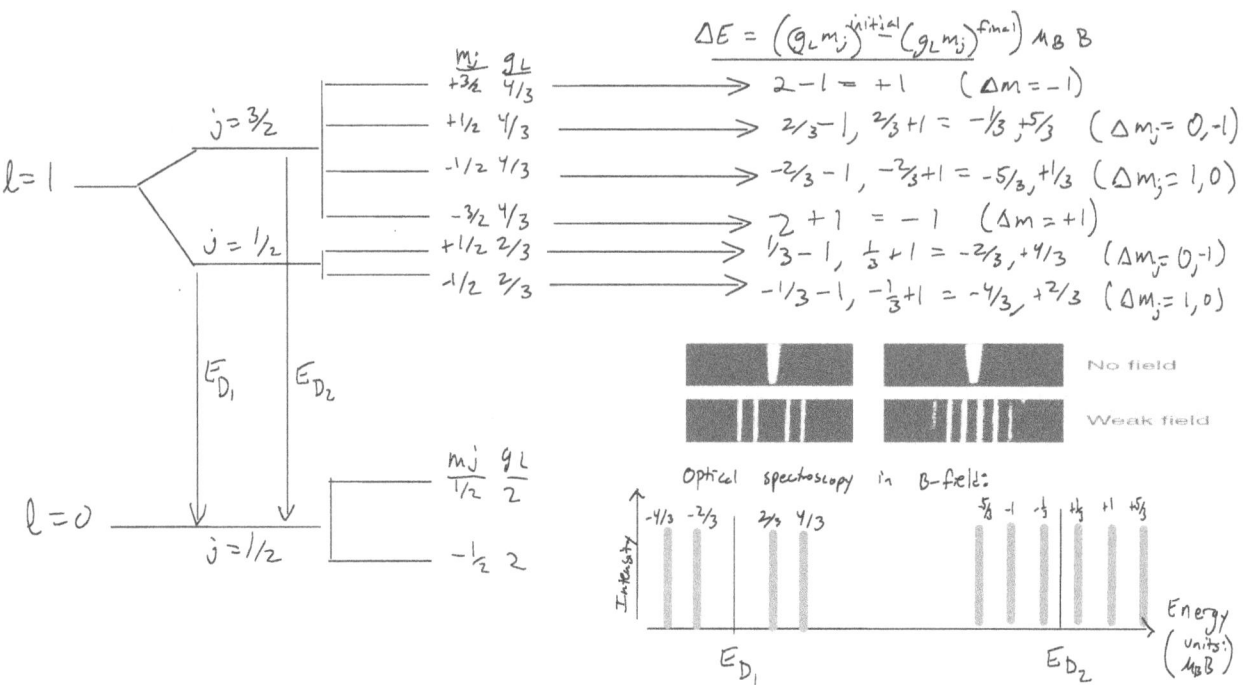

$$\Delta E = \left((g_L m_j)^{initial} - (g_L m_j)^{final}\right) \mu_B B$$

Nuclear angular momentum: hyperfine interaction

The nucleons also have angular momentum and magnetic moment. This couples to the electron spin by "stray" B-field:
Classically, the B-field of a magnetic dipole is

$$\vec{B} = \frac{\mu_0}{4\pi r^3}\left[3(\vec{\mu}\cdot\hat{r})\hat{r} - \vec{\mu}\right] + \frac{2\mu_0}{3}\vec{\mu}\,\delta^3(\vec{r}),$$

with proton moment $\vec{\mu}_p = g_p \frac{\mu_e}{\hbar}\vec{S}_p$. This contributes a term to the electron Hamiltonian:

$$-\vec{\mu}_e \cdot \vec{B} = g_e \frac{\mu_B}{\hbar}\vec{S}_e \cdot \vec{B} = \frac{\mu_0}{\hbar}g_e g_p \mu_B \mu_p \left[\frac{3(\vec{S}_p\cdot\hat{r})(\vec{S}_e\cdot\hat{r}) - \vec{S}_p\cdot\vec{S}_e}{4\pi r^3} + \underbrace{\frac{2}{3}\vec{S}_p\cdot\vec{S}_e\,\delta^3(\vec{r})}_{\text{"Contact term"}}\right]$$

For $\ell=0$ (spherically symmetric) states, the expectation value of 1st term is zero, but the contact term picks out $|\Psi|^2$ @ origin $\sim \frac{1}{a_0^3}$ (same as our approx. for $\sim \langle \frac{1}{r^3}\rangle$).

$\mu_B = \frac{e\hbar}{2m_e}$ so $\mu_p = \frac{e\hbar}{2m_p} = \mu_B\left(\frac{m_e}{m_p}\right)$ and $E_{hyperfine} \sim \left(\frac{m_e}{m_p}\right)\frac{\mu_B^2}{a_0^3 \epsilon_0 c^2} = \left(\frac{m_e}{m_p}\right)\alpha^4 E_0 \sim \text{meV}$

Time-independent perturbation theory

Our approach to understanding the Zeeman + SOI spectrum vs. B-field can be mathematically generalized in the following way:

If we can diagonalize a simple Hamiltonian exactly: $H^0|\varphi_n^0\rangle = E_n^0|\varphi_n^0\rangle$ can we approximately diagonalize a similar Hamiltonian $(H^0 + \lambda H')|\varphi_n\rangle = E_n|\varphi_n\rangle$, where $\langle \lambda H'\rangle \ll \langle H^0 \rangle$??

(e.g. $\lambda H' \to$ Zeeman for weak B field, $H^0 \to$ SOI, OR $\lambda H' \to$ SOI, $H^0 \to$ Zeeman for strong field).

We expect that eigenvalues and eigenfunctions can be expanded in powers of λ: $|\varphi_n\rangle = |\varphi_n^0\rangle + \lambda|\varphi_n^1\rangle + \lambda^2|\varphi_n^2\rangle + \ldots$, $E_n = E_n^0 + \lambda E_n^1 + \lambda^2 E_n^2 + \ldots$

Then, by direct substitution:

$$(H^0 + \lambda H')(|\varphi_n^0\rangle + \lambda|\varphi_n^1\rangle + \lambda^2|\varphi_n^2\rangle + \ldots) = (E_n^0 + \lambda E_n^1 + \lambda^2 E_n^2 + \ldots)(|\varphi_n^0\rangle + \lambda|\varphi_n^1\rangle + \lambda^2|\varphi_n^2\rangle + \ldots)$$

Now, collect terms to the same power of λ on each side.

To zeroth order in λ, we have our solved system $H^0|\varphi_n^0\rangle = E_n^0|\varphi_n^0\rangle$.

First-order correction to energy eigenvalues

To **first** order in λ: $\quad H^0|\varphi_n^1\rangle + H'|\varphi_n^0\rangle = E_n^1|\varphi_n^0\rangle + E_n^0|\varphi_n^1\rangle$

Project onto known zeroth-order eigenfunctions:

$\langle \varphi_n^0| \times \implies \langle \varphi_n^0|H^0|\varphi_n^1\rangle + \langle \varphi_n^0|H'|\varphi_n^0\rangle = \langle \varphi_n^0|E_n^1|\varphi_n^0\rangle + \cancel{\langle \varphi_n^0|E_n^0|\varphi_n^1\rangle}$

Exploit hermiticity of H^0:

$\langle \varphi_n^0|H^0 = (H^{0\dagger}|\varphi_n^0\rangle)^\dagger = (H^0|\varphi_n^0\rangle)^\dagger = (E_n^0|\varphi_n^0\rangle)^\dagger = \langle \varphi_n^0|E_n^{0*} = \langle \varphi_n^0|E_n^0$

So $\langle \varphi_n^0|H^0|\varphi_n^1\rangle = \langle \varphi_n^0|E_n^0|\varphi_n^1\rangle$, and cancellation of two terms above.

This leaves $\langle \varphi_n^0|E_n^1|\varphi_n^0\rangle = E_n^1\langle \varphi_n^0|\varphi_n^0\rangle = \boxed{E_n^1 = \langle \varphi_n^0|H'|\varphi_n^0\rangle}$

So, we assume that the perturbation leaves wavefunctions unchanged, and calculate the expectation values (diagonal matrix elements) → we've been doing this all along.

But, the perturbation **does** change the wavefunction (in general). How? We can calculate it!

A note on completeness

If the set of $|\psi_n^0\rangle$ are orthonormal and complete, then we can expand any arbitrary state in this basis:

$$|\phi\rangle = c_1|\psi_1^0\rangle + c_2|\psi_2^0\rangle + c_3|\psi_3^0\rangle + \ldots$$

Let's see what happens when we multiply by this operator:

$$\sum_n |\psi_n^0\rangle\langle\psi_n^0|\phi\rangle = \sum_n |\psi_n^0\rangle \left(c_1 \langle\psi_n^0|\psi_1^0\rangle + c_2 \langle\psi_n^0|\psi_2^0\rangle + \ldots \right)$$

$$= \sum_n |\psi_n^0\rangle \left(c_1 \delta_{n1} + c_2 \delta_{n2} + \ldots \right)$$

$$= c_1|\psi_1^0\rangle + c_2|\psi_2^0\rangle + c_3|\psi_3^0\rangle + \ldots = |\phi\rangle$$

In other words, for any complete orthonormal basis,

$$\sum_n |\psi_n\rangle\langle\psi_n| = \mathbb{1} \quad \text{(identity operator)}$$

First-order correction to wavefunction

Start again with terms of Schrödinger eqn proportional to λ:

$$H'|\psi_n^0\rangle + H^0|\psi_n'\rangle = E_n^0|\psi_n'\rangle + E_n'|\psi_n^0\rangle$$

Rewrite first term:

$$H'|\psi_n^0\rangle = \underbrace{\sum_k |\psi_k^0\rangle\langle\psi_k^0|}_{=\mathbb{1} \text{ due to completeness}} H'|\psi_n^0\rangle = \sum_{k\neq n} |\psi_k^0\rangle\langle\psi_k^0|H'|\psi_n^0\rangle + |\psi_n^0\rangle\langle\psi_n^0|H'|\psi_n^0\rangle$$

$$= \sum_{k\neq n} |\psi_k^0\rangle\langle\psi_k^0|H'|\psi_n^0\rangle + E_n'|\psi_n^0\rangle$$

Now substitute:

$$\sum_{k\neq n} |\psi_k^0\rangle\langle\psi_k^0|H'|\psi_n^0\rangle + \cancel{E_n'|\psi_n^0\rangle} + H^0|\psi_n'\rangle = E_n^0|\psi_n'\rangle + \cancel{E_n'|\psi_n^0\rangle}$$

$$(E_n^0 - H^0)|\psi_n'\rangle = \sum_{k\neq n} |\psi_k^0\rangle\langle\psi_k^0|H'|\psi_n^0\rangle$$

and project onto unperturbed states:

$$\langle\psi_\ell^0| \times \quad \longrightarrow \quad \langle\psi_\ell^0|(E_n^0 - H^0)|\psi_n'\rangle = \sum_{k\neq n} \underbrace{\langle\psi_\ell^0|\psi_k^0\rangle}_{\delta_{\ell k} \text{ due to orthonormality}} \langle\psi_k^0|H'|\psi_n^0\rangle = \langle\psi_\ell^0|H'|\psi_n^0\rangle$$

Projection onto unperturbed eigenstates

Thanks to hermiticity of H^0, we have

$$(E_n^0 - E_\ell^0)\langle \psi_\ell^0 | \psi_n^1 \rangle = \langle \psi_\ell^0 | H' | \psi_n^0 \rangle \implies \langle \psi_\ell^0 | \psi_n^1 \rangle = \frac{\langle \psi_\ell^0 | H' | \psi_n^0 \rangle}{E_n^0 - E_\ell^0}$$

This inner product gives the coefficients in a projection of the unknown $|\psi_n^1\rangle$ onto the known orthonormal set of $|\psi_\ell^0\rangle$!

To see this explicitly, take outer products and sum:

$$\underbrace{\sum_\ell |\psi_\ell^0\rangle \langle \psi_\ell^0 | \psi_n^1 \rangle}_{=\mathbb{1} \text{ by completeness}} = \sum_\ell |\psi_\ell^0\rangle \frac{\langle \psi_\ell^0 | H' | \psi_n^0 \rangle}{E_n^0 - E_\ell^0}$$

$$\boxed{|\psi_n^1\rangle = \sum_{\ell \neq n} \frac{\langle \psi_\ell^0 | H' | \psi_n^0 \rangle}{E_n^0 - E_\ell^0} |\psi_\ell^0\rangle}$$

So the wavefunction correction is determined by scaled off-diagonal elements of the perturbation Hamiltonian. Notice that, due to the denominator, this is valid only in the non-degenerate case $E_n^0 \neq E_\ell^0$ for any ℓ.

Second-order correction to eigenvalues

Go back to Schrödinger equation, and select terms proportional to λ^2:

$$H'|\psi_n^1\rangle + H^0|\psi_n^2\rangle = E_n^1|\psi_n^1\rangle + E_n^2|\psi_n^0\rangle + E_n^0|\psi_n^2\rangle$$

Now, project onto unperturbed basis:

$$\langle \psi_n^0 | \times \implies \langle \psi_n^0 | H' | \psi_n^1 \rangle + \cancel{\langle \psi_n^0 | H^0 | \psi_n^2 \rangle} = \underbrace{\langle \psi_n^0 | E_n^1 | \psi_n^1 \rangle}_{=0 \text{ because } |\psi_n^1\rangle \text{ has no } |\psi_n^0\rangle \text{ component (remember } \ell \neq n \text{ in sum)}} + \langle \psi_n^0 | E_n^2 | \psi_n^0 \rangle + \cancel{\langle \psi_n^0 | E_n^0 | \psi_n^2 \rangle}$$

(cancel due to hermiticity of H^0)

So we have $E_n^2 = \langle \psi_n^0 | H' | \psi_n^1 \rangle$. Now substitute 1st order correction to wavefunction:

$$E_n^2 = \langle \psi_n^0 | H' \left(\sum_{\ell \neq n} \frac{\langle \psi_\ell^0 | H' | \psi_n^0 \rangle}{E_n^0 - E_\ell^0} |\psi_\ell^0\rangle \right) = \boxed{\sum_{\ell \neq n} \frac{|\langle \psi_\ell^0 | H' | \psi_n^0 \rangle|^2}{E_n^0 - E_\ell^0}}$$

Again, this only works for non-degenerate E_n^0! We can use this result to calculate the 2nd-order wavefunction, and then 3rd order eigenvalue E_n^3, etc, bootstrapping our way to ever-higher orders.

Numerical results

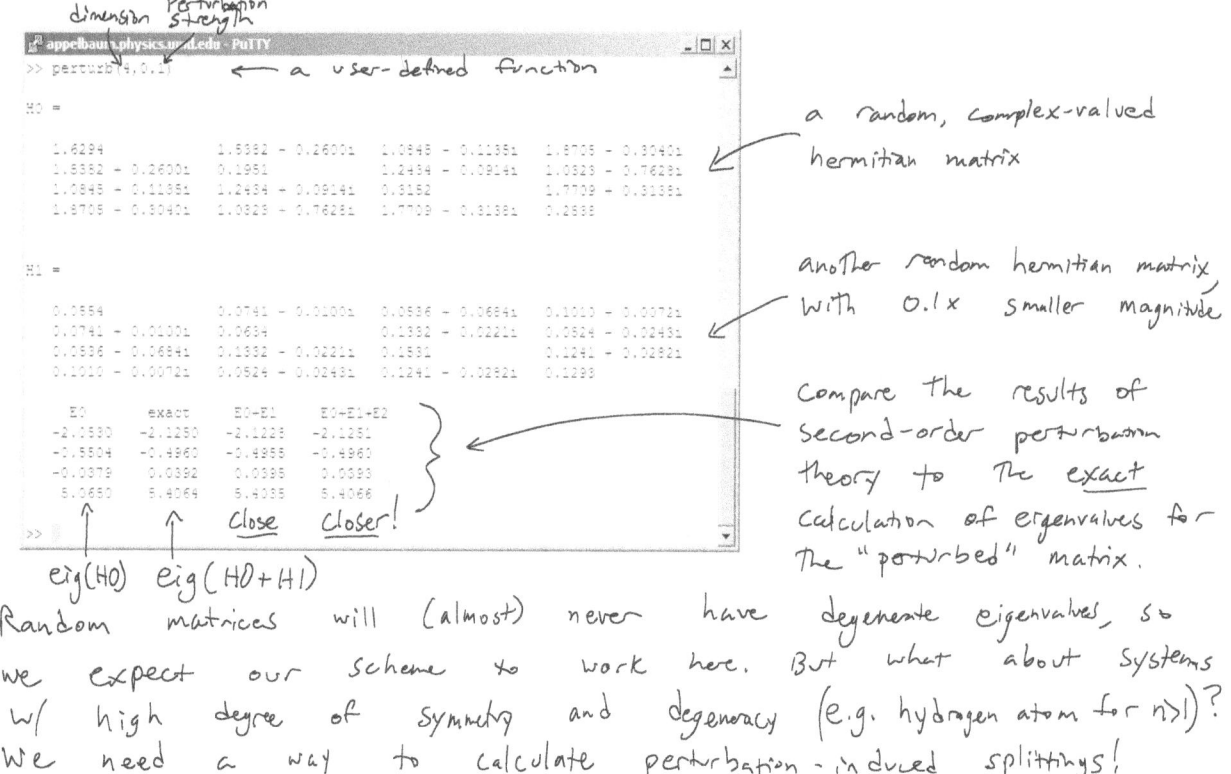

Random matrices will (almost) never have degenerate eigenvalues, so we expect our scheme to work here. But what about systems w/ high degree of symmetry and degeneracy (e.g. hydrogen atom for n>1)? We need a way to calculate perturbation-induced splittings!

A tractable example, by hand

Our unperturbed Hamiltonian $H^0 = \eta \begin{bmatrix} 0 & i \\ 1 & 0 \end{bmatrix}$ has eigenvalues and eigenvectors:

$$E_1^0 = +\eta, \quad |\psi_1^0\rangle = \frac{1}{\sqrt{2}}\begin{bmatrix} 1 \\ 1 \end{bmatrix} \quad \text{and} \quad E_2^0 = -\eta, \quad |\psi_2^0\rangle = \frac{1}{\sqrt{2}}\begin{bmatrix} 1 \\ -1 \end{bmatrix}$$

What are the corrections to eigenvalues caused by the perturbation $H' = \lambda \begin{bmatrix} 0 & -i \\ i & 0 \end{bmatrix}$?

1st order:
$$E_1^1 = \langle \psi_1^0 | H' | \psi_1^0 \rangle = \frac{1}{\sqrt{2}}[1\ 1]\lambda\begin{bmatrix} 0 & -i \\ i & 0 \end{bmatrix}\frac{1}{\sqrt{2}}\begin{bmatrix} 1 \\ 1 \end{bmatrix} = \frac{\lambda}{2}[1\ 1]\begin{bmatrix} -i \\ i \end{bmatrix} = 0$$

$$E_2^1 = \langle \psi_2^0 | H' | \psi_2^0 \rangle = \frac{1}{\sqrt{2}}[1\ -1]\lambda\begin{bmatrix} 0 & -i \\ i & 0 \end{bmatrix}\frac{1}{\sqrt{2}}\begin{bmatrix} 1 \\ -1 \end{bmatrix} = \frac{\lambda}{2}[1\ -1]\begin{bmatrix} i \\ i \end{bmatrix} = 0$$

Not surprising, since our $|\psi_n^0\rangle$ are spins polarized along $+x$ and $-x$, and the perturbation is a field along y ($E = -\vec{\mu}\cdot\vec{B} = 0$)!

2nd order: $E_n^2 = \sum_{\ell \neq n} \frac{|\langle \psi_\ell^0 | H' | \psi_n^0 \rangle|^2}{E_n - E_\ell}$. For 2x2, only one term: $E_1^2 = \frac{|\langle \psi_1^0 | H' | \psi_2^0 \rangle|^2}{E_1 - E_2}$

$$E_1^2 = \frac{\left|\frac{\lambda}{2}[1\ 1]\begin{bmatrix} 0 & -i \\ i & 0 \end{bmatrix}\begin{bmatrix} 1 \\ -1 \end{bmatrix}\right|^2}{\eta - (-\eta)} = \frac{\lambda^2}{2\eta}, \quad E_2^2 = \frac{|\langle \psi_2^0 | H' | \psi_1^0 \rangle|^2}{E_2 - E_1} = -E_1^2 = -\frac{\lambda^2}{2\eta}$$

So our approximate energy eigenvalues are $\pm \eta \pm \frac{\lambda^2}{2\eta}$.

Comparison to exact diagonalization

$H^0 + H' = \begin{bmatrix} 0 & \eta - i\lambda \\ \eta + i\lambda & 0 \end{bmatrix}$, characteristic polynomial given by: $\det \begin{vmatrix} -E & \eta - i\lambda \\ \eta + i\lambda & -E \end{vmatrix} = 0$

Find roots of $E^2 - (\eta^2 + \lambda^2) = 0$, with solutions $E = \pm\sqrt{\eta^2 + \lambda^2}$.

If $\lambda \ll \eta$, we can expand in a power series

$\pm\sqrt{\eta^2 + \lambda^2} = \pm\eta\sqrt{1 + \frac{\lambda^2}{\eta^2}} = \pm\eta\left(1 + \frac{\lambda^2}{2\eta^2} + \ldots\right) \approx \pm\eta \pm \frac{\lambda^2}{2\eta}$ (This is the same as our 2nd order perturbation theory result!)

[Figure: Energy diagram showing exact hyperbolic curves and 2nd order perturbation theory approximation vs λ, with asymptotic eigenstates $\frac{1}{\sqrt{2}}(|\uparrow\rangle - i|\downarrow\rangle)$, $\frac{1}{\sqrt{2}}(|\uparrow\rangle + i|\downarrow\rangle)$ labeled at branches; eigenstates $\frac{1}{\sqrt{2}}(|\uparrow\rangle + |\downarrow\rangle)$ at $+\eta$ and $\frac{1}{\sqrt{2}}(|\uparrow\rangle - |\downarrow\rangle)$ at $-\eta$; curves labeled "exact" and "2nd order pert. Thry: $E^0 + E^1 + E^2$"; asymptotic eigenstate of σ_y]

A perspective into degeneracy

We still need to figure out how to apply perturbation theory to degenerate states. Lets look at <u>why</u> it fails and try to circumvent the problem.

Suppose we can diagonalize a 2×2 Hamiltonian H^0 and obtain eigenvalues and eigenvectors $E_1^0, |\psi_1^0\rangle$ and $E_2^0, |\psi_2^0\rangle$.

What are the eigenvalues of the perturbed Hamiltonian $H^0 + H'$? In the <u>nearly</u> diagonalizing basis $|\psi_1^0\rangle$ and $|\psi_2^0\rangle$, we have

$$\begin{bmatrix} \langle \psi_1^0 | H^0 + H' | \psi_1^0 \rangle & \langle \psi_1^0 | H^0 + H' | \psi_2^0 \rangle \\ \langle \psi_2^0 | H^0 + H' | \psi_1^0 \rangle & \langle \psi_2^0 | H^0 + H' | \psi_2^0 \rangle \end{bmatrix} = \begin{bmatrix} E_1^0 + E_1^1 & \Delta \\ \Delta^* & E_2^0 + E_2^1 \end{bmatrix}$$

where $E_1^1 = \langle \psi_1^0 | H' | \psi_1^0 \rangle$ and $E_2^1 = \langle \psi_2^0 | H' | \psi_2^0 \rangle$, and $\Delta = \langle \psi_1^0 | H' | \psi_2^0 \rangle$. First-order perturbation theory would just take the diagonal elements and neglect Δ entirely.

Approximating eigenvalues

The exact eigenvalues are given by the roots of

$$\begin{vmatrix} (E_1^0 + E_1^1) - E & \Delta \\ \Delta^* & (E_2^0 + E_2^1) - E \end{vmatrix} = 0$$

Writing $E_1^* = E_1^0 + E_1^1$ and $E_2^* = E_2^0 + E_2^1$,

$$(E_1^* - E)(E_2^* - E) - |\Delta|^2 = 0 \longrightarrow E^2 - (E_1^* + E_2^*)E + E_1^* E_2^* - |\Delta|^2 = 0$$

By quadratic formula,

$$E = \frac{E_1^* + E_2^* \pm \sqrt{(E_1^* + E_2^*)^2 - 4E_1^* E_2^* + 4|\Delta|^2}}{2} = \frac{E_1^* + E_2^* \pm \sqrt{(E_1^* - E_2^*)^2 + 4|\Delta|^2}}{2}$$

Now, if $|\Delta|^2 \ll (E_1^* - E_2^*)^2$, it makes sense to expand

$$\frac{E_1^* + E_2^* \pm (E_1^* - E_2^*)\sqrt{1 + \frac{4|\Delta|^2}{(E_1^* - E_2^*)^2}}}{2} = \frac{E_1^* + E_2^* \pm (E_1^* - E_2^*)\left(1 + 2\frac{|\Delta|^2}{(E_1^* - E_2^*)^2} + \ldots\right)}{2}$$

$$\approx \begin{cases} (+): E_1^* + \frac{|\Delta|^2}{E_1^* - E_2^*} \\ (-): E_2^* + \frac{|\Delta|^2}{E_2^* - E_1^*} \end{cases}$$ Compare with 2nd order perturbation theory!: $E_n \sim E_n^* + \sum_{\ell \neq n} \frac{|\Delta|^2}{E_n^0 - E_\ell^0}$

Failure under degeneracy

If $E_1 = E_2$, then 2nd order result diverges due to vanishing denominator.

- The problem is caused by an invalid expansion of $\frac{E_1 + E_2 \pm \sqrt{(E_1 - E_2)^2 + 4|\Delta|^2}}{2}$ which assumed that $|\Delta|^2 \ll (E_1 - E_2)^2$, not true for degenerate states.

- If $E_1 = E_2$, then <u>without</u> any approximation or Taylor expansion, $E = E_{1,2}^{(1)} \pm |\Delta|$ and the degeneracy is broken by a <u>finite</u> amount.

- So our shortcut won't work for degenerate states, and we have to go back closer to the beginning: <u>directly diagonalize</u> H' in the degenerate subspace for first-order corrections to the energy, rather than just taking diagonal elements of $\langle \varphi_n^0 | H' | \varphi_m^0 \rangle$.

A tractable example, by hand

$$H_0 = \eta \begin{bmatrix} 0 & 0 & 0 \\ 0 & 1 & 1 \\ 0 & 1 & 1 \end{bmatrix}$$

eigenvalues determined by roots of
$$-E((\eta-E)^2 - \eta^2) = 0$$

$$\begin{cases} E_1^0: 0, |\psi_1^0\rangle = \begin{bmatrix} 1 \\ 0 \\ 0 \end{bmatrix} \\ E_2^0: 0, |\psi_2^0\rangle = \frac{1}{\sqrt{2}}\begin{bmatrix} 0 \\ -1 \\ 1 \end{bmatrix} \\ E_3^0: 2\eta, |\psi_3^0\rangle = \frac{1}{\sqrt{2}}\begin{bmatrix} 0 \\ 1 \\ 1 \end{bmatrix} \end{cases}$$

Our perturbation $H' = \lambda \begin{bmatrix} 0 & 0 & 1 \\ 0 & 0 & 0 \\ 1 & 0 & 0 \end{bmatrix}$. Blindly applying <u>nondegenerate</u> perturbation theory,

$E_3^1: \langle \psi_3^0 | H' | \psi_3^0 \rangle = \frac{\lambda}{2} \begin{bmatrix} 0 & 1 & 1 \end{bmatrix} \begin{bmatrix} 0 & 0 & 1 \\ 0 & 0 & 0 \\ 1 & 0 & 0 \end{bmatrix} \begin{bmatrix} 0 \\ 1 \\ 1 \end{bmatrix} = 0$ ok!

$E_2^1: \langle \psi_2^0 | H' | \psi_2^0 \rangle = \frac{\lambda}{2} \begin{bmatrix} 0 & -1 & 1 \end{bmatrix} \begin{bmatrix} 0 & 0 & 1 \\ 0 & 0 & 0 \\ 1 & 0 & 0 \end{bmatrix} \begin{bmatrix} 0 \\ -1 \\ 1 \end{bmatrix} = 0$

$E_1^1: \langle \psi_1^0 | H' | \psi_1^0 \rangle = \lambda \begin{bmatrix} 1 & 0 & 0 \end{bmatrix} \begin{bmatrix} 0 & 0 & 1 \\ 0 & 0 & 0 \\ 1 & 0 & 0 \end{bmatrix} \begin{bmatrix} 1 \\ 0 \\ 0 \end{bmatrix} = 0$

wrong, since these unperturbed states are degenerate!

Degenerate perturbation theory:
Instead of taking just the diagonal elements, take eigenvalues of diagonal block of degenerate subspace:

$$\begin{bmatrix} \langle \psi_1^0 | H' | \psi_1^0 \rangle & \langle \psi_1^0 | H' | \psi_2^0 \rangle \\ \langle \psi_2^0 | H' | \psi_1^0 \rangle & \langle \psi_2^0 | H' | \psi_2^0 \rangle \end{bmatrix} = \begin{bmatrix} 0 & \frac{\lambda}{\sqrt{2}} \\ \frac{\lambda}{\sqrt{2}} & 0 \end{bmatrix} \Rightarrow E_{1,2}^1 = \pm \frac{\lambda}{\sqrt{2}}$$

Second order degenerate

The perturbation splits E_1 and E_2 by $\pm \frac{\lambda}{\sqrt{2}}$, and also mixes ψ_1 and ψ_2. Thus to go to higher order we must use basis $\{\frac{1}{\sqrt{2}}(|\psi_1\rangle \pm |\psi_2\rangle), |\psi_3\rangle\}$

In this basis, H' is

$$\lambda \begin{bmatrix} \frac{1}{\sqrt{2}} & -\frac{1}{2} & \frac{1}{2} \\ \frac{1}{\sqrt{2}} & \frac{1}{2} & -\frac{1}{2} \\ 0 & \frac{1}{\sqrt{2}} & \frac{1}{\sqrt{2}} \end{bmatrix} \begin{bmatrix} 0 & 0 & 1 \\ 0 & 0 & 0 \\ 1 & 0 & 0 \end{bmatrix} \begin{bmatrix} \frac{1}{\sqrt{2}} & \frac{1}{\sqrt{2}} & 0 \\ -\frac{1}{2} & \frac{1}{2} & \frac{1}{\sqrt{2}} \\ \frac{1}{2} & -\frac{1}{2} & \frac{1}{\sqrt{2}} \end{bmatrix} = \lambda \begin{bmatrix} \frac{1}{\sqrt{2}} & 0 & \frac{1}{2} \\ 0 & -\frac{1}{\sqrt{2}} & \frac{1}{2} \\ \frac{1}{2} & \frac{1}{2} & 0 \end{bmatrix}$$

So 2nd order perturbation theory gives:

$E_1 = 0 + \frac{\lambda}{\sqrt{2}} + \frac{(\lambda/2)^2}{0 - 2\eta} = \frac{\lambda}{\sqrt{2}} - \frac{\lambda^2}{8\eta}$

$E_2 = 0 - \frac{\lambda}{\sqrt{2}} + \frac{(\lambda/2)^2}{0 - 2\eta} = -\frac{\lambda}{\sqrt{2}} - \frac{\lambda^2}{8\eta}$

$E_3 = 2\eta + 0 + \frac{(\lambda/2)^2}{2\eta - 0} + \frac{(\lambda/2)^2}{2\eta - 0} = 2\eta + \frac{\lambda^2}{4\eta}$

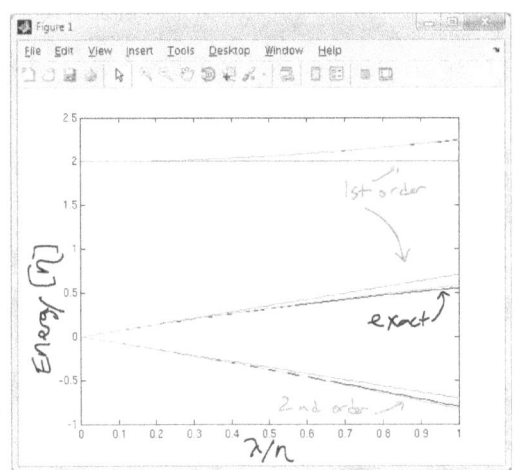

Example: Non-degenerate perturbation theory

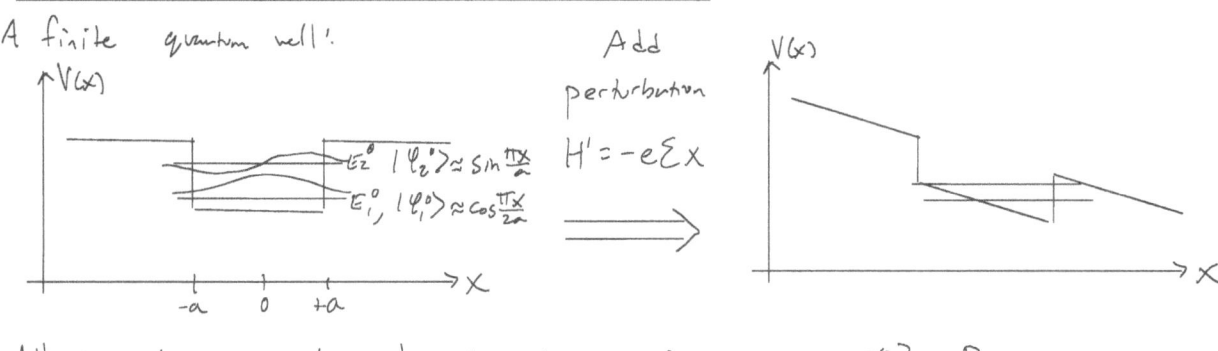

A finite quantum well. Add perturbation $H' = -e\mathcal{E}x$

What happens to bound state w/ energy E_1^0? Perturbation theory:

$$E_1 = E_1^0 + \langle \psi_1^0 | H' | \psi_1^0 \rangle + \sum_{\ell \neq 1} \frac{|\langle \psi_1^0 | H' | \psi_n^0 \rangle|^2}{E_1 - E_\ell} + \cdots$$

1st order: $\langle \psi_1^0 | H' | \psi_1^0 \rangle \approx \frac{1}{a} \int_{-a}^{+a} \overset{\text{even}}{\cos \frac{\pi x}{2a}} \overset{\text{odd}}{(-e\mathcal{E}x)} \overset{\text{even}}{\cos \frac{\pi x}{2a}} dx = 0$ by symmetry!

2nd order: $\dfrac{\overset{\text{positive, nonzero}}{|\langle \psi_1^0 | H' | \psi_2^0 \rangle|^2}}{\underset{\text{negative}}{E_1^0 - E_2^0}} < 0 \longrightarrow$ ground state decreases in energy proportional to $\mathcal{E}^2$!

Quantum-confined Stark effect

$$\frac{|\langle \psi_2^0 | -e\mathcal{E}x | \psi_1^0 \rangle|^2}{E_1^0 - E_2^0} \approx \frac{e^2\mathcal{E}^2 \left(\frac{1}{a} \int_{-a}^{a} \sin \frac{\pi x}{a} \cdot x \cdot \cos \frac{\pi x}{2a} dx \right)^2}{\frac{\hbar^2 \pi^2}{2m(2a)^2}(1-4)} \longleftarrow \text{infinite QW energies}$$

$\frac{1}{a} \int_{-a}^{a} \sin \frac{\pi x}{a} \cdot x \cdot \cos \frac{\pi x}{2a} dx = \frac{2}{a} \int_{-a}^{a} \sin \frac{\pi x}{2a} \cos^2 \frac{\pi x}{2a} x \, dx$. ($\sin 2x = 2\sin x \cos x$)

integration by parts ($\int u\,dv = uv - \int v\,du$) $u = x$, $dv = \cos^2 \frac{\pi x}{2a} \sin \frac{\pi x}{2a} dx$
$du = dx$, $v = -\frac{\cos^3 \frac{\pi x}{2a}}{3} \cdot \frac{2a}{\pi}$

$\frac{2ax}{3\pi} \cos^3 \frac{\pi x}{2a} \Big|_{-a}^{a} \overset{0}{} - \frac{2a}{3\pi} \int_{-a}^{a} \cos^3 \frac{\pi x}{2a} dx = -\frac{2a}{3\pi} \int_{-a}^{a} \cos \frac{\pi x}{2a} (1 - \sin^2 \frac{\pi x}{2a}) dx$

$= -\frac{2a}{3\pi} \cdot \frac{2a}{\pi} \left(\sin \frac{\pi x}{2a} - \frac{\sin^3 \frac{\pi x}{2a}}{3} \right) \Big|_{-a}^{a} = -\frac{4a^2}{3\pi^2} \left(\frac{2}{3} - \left(-\frac{2}{3}\right) \right) = \left(\frac{2(2a)}{3\pi} \right)^2$

So our 2nd-order correction is   (L = 2a)

$$\frac{|\langle \psi_2^0 | -e\mathcal{E}x | \psi_1^0 \rangle|^2}{E_1 - E_2} \approx -\left(\frac{4e\mathcal{E}}{2a}\right)^2 \frac{2m\left(\frac{2(2a)}{3\pi}\right)^4}{3\hbar^2 \pi^2} = -24 \cdot \left(\frac{2}{3\pi}\right)^6 \frac{e^2\mathcal{E}^2 m L^4}{\hbar^2}$$

This mechanism enables the high-frequency modulation of solid-state laser gain for optical fiber telecom.

118

Stark effect in Hydrogen

$H^0 = -\frac{\hbar^2}{2m}\nabla^2 - \frac{e^2}{r}$. Due to spherical symmetry, the Hamiltonian is separable in r, θ, ϕ coordinates and hence $|\psi^0_{n\ell m}\rangle = R_{n\ell}(r) Y_\ell^m(\theta,\phi)$.

What happens to energy spectrum under perturbation $e\mathcal{E}z = e\mathcal{E}|r|\cos\theta$?

- **ground state:** since dipole $\vec{p} = \langle \psi^0_{100}|\vec{r}|\psi^0_{100}\rangle = 0$ for unperturbed state, we expect nonzero correction only in second order (just like in 1-D).

- **excited state, n=2:** 4 degenerate states $|\psi^0_{200}\rangle, |\psi^0_{21-1}\rangle, |\psi^0_{210}\rangle, |\psi^0_{211}\rangle$

Use degenerate perturbation theory to calculate first-order corrections! Construct the perturbation operator H' in the degenerate subspace. 16 elements total:

$$e\mathcal{E}\begin{bmatrix} \langle\psi^0_{200}|z|\psi^0_{200}\rangle & \langle\psi^0_{200}|z|\psi^0_{21-1}\rangle & \langle\psi^0_{200}|z|\psi^0_{210}\rangle & \langle\psi^0_{200}|z|\psi^0_{211}\rangle \\ \langle\psi^0_{21-1}|z|\psi^0_{200}\rangle & \langle\psi^0_{21-1}|z|\psi^0_{21-1}\rangle & \langle\psi^0_{21-1}|z|\psi^0_{210}\rangle & \langle\psi^0_{21-1}|z|\psi^0_{211}\rangle \\ \langle\psi^0_{210}|z|\psi^0_{200}\rangle & \langle\psi^0_{210}|z|\psi^0_{21-1}\rangle & \langle\psi^0_{21-1}|z|\psi^0_{21-1}\rangle & \langle\psi^0_{211}|z|\psi^0_{211}\rangle \\ \langle\psi^0_{211}|z|\psi^0_{200}\rangle & \langle\psi^0_{211}|z|\psi^0_{21-1}\rangle & \langle\psi^0_{21-1}|z|\psi^0_{211}\rangle & \langle\psi^0_{211}|z|\psi^0_{211}\rangle \end{bmatrix}$$

Exploiting symmetry

- Due to Hermiticity, only 10 matrix elements are unique.
- Due to inversion symmetry of $\psi^*\psi$, all unperturbed states have no dipole moment so all diagonal elements are zero.
- Operator $z = |r|\cos\theta$ does not affect integration over ϕ. Azimuthal part of Y_ℓ^m is $e^{im\phi}$ so matrix elements are proportional to
$$\int_0^{2\pi} e^{-im'\phi} e^{im\phi} d\phi = 0 \quad \text{unless } m=m'$$

Only nonzero matrix element is between states of same m: $\gamma = \langle\psi^0_{200}|z|\psi^0_{210}\rangle$

Thus, our degenerate subspace Hamiltonian is vastly easier to diagonalize:

$$H' = e\mathcal{E}\begin{bmatrix} 0 & 0 & \gamma & 0 \\ 0 & 0 & 0 & 0 \\ \gamma^* & 0 & 0 & 0 \\ 0 & 0 & 0 & 0 \end{bmatrix}$$

Characteristic polynomial

$$\det \begin{vmatrix} -E' & 0 & e\mathcal{E}\gamma & 0 \\ 0 & -E' & 0 & 0 \\ e\mathcal{E}\gamma^* & 0 & -E' & 0 \\ 0 & 0 & 0 & -E' \end{vmatrix} = 0$$

$E'^4 - e^2\mathcal{E}^2\gamma|^2 E'^2 = 0 \longrightarrow$ roots are $E' = 0, 0, \pm e\mathcal{E}|\gamma|$

$E' = 0$ eigenvalues correspond to $|\psi_{21\pm1}\rangle$ states.

$E' = \pm e\mathcal{E}\gamma$ eigenvalues correspond to $\frac{1}{\sqrt{2}}(|\psi_{200}\rangle \pm |\psi_{210}\rangle)$ superpositions.

$\gamma = \langle \psi_{200}|r\cos\theta|\psi_{210}\rangle = \int_0^\infty r^3 dr \left[\frac{1}{(2a_0)^{3/2}}(2-\frac{r}{a_0})e^{-\frac{r}{2a_0}}\right]\left[\frac{1}{(2a_0)^{3/2}}\frac{1}{\sqrt{3}}\frac{r}{a_0}e^{-\frac{r}{2a_0}}\right]\underbrace{\int d\Omega\, Y_{00}^*\cos\theta\, Y_{10}}_{\frac{1}{\sqrt{3}}}$

$= \frac{1}{\sqrt{3}}\int_0^\infty \frac{(2-\frac{r}{a_0})}{8\sqrt{3}\,a_0^4} e^{-\frac{r}{a_0}} r^4 dr = \frac{1}{24 a_0^4}\int_0^\infty (2r^4 - \frac{r^5}{a_0})e^{-\frac{r}{a_0}}dr$

If $x \equiv \frac{r}{a_0}$, $= \frac{a_0}{24}\int_0^\infty (2x^4 - x^5)e^{-x}dx$

Using $\int_0^\infty x^n e^{-x}dx = n!$, $= \frac{a_0}{24}(2 \cdot 24 - 120) = -3a_0$. So $\Delta E = \pm 3e\mathcal{E}a_0$ and 0

Periodic potentials

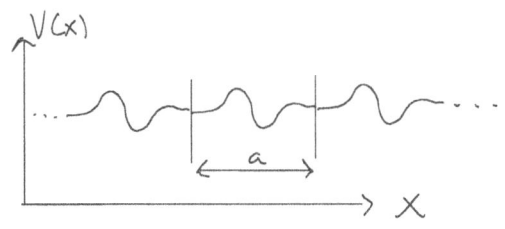

- Solve time-independent Schrödinger equation $\left(-\frac{\hbar^2}{2m}\frac{\partial^2}{\partial x^2} + V(x)\right)\psi = E\psi$ for $V(x+a) = V(x)$ "periodic translational symmetry"

- If $V(x) = $ const., V has "continuous translational symmetry", and we know $\psi(x) = e^{ikx}$, (planewave states labeled by k, w/ "dispersion relation" $E = \frac{\hbar^2 k^2}{2m}$).

- If $V(x) \neq $ const., similarly $\psi(x) = u(x)e^{ikx}$, where $u(x) = u(x+a)$, called "Bloch wave"

k-dot-p perturbation

- Substitution into T.I.S.E. gives

$$-\frac{\hbar^2}{2m}(ue^{ikx})'' + Vue^{ikx} = Eue^{ikx}$$

$$-\frac{\hbar^2}{2m}(u'' + 2u'\cdot ik - k^2 u)e^{ikx} + Vue^{ikx} = Eue^{ikx}$$

$$\left[\underbrace{\left(-\frac{\hbar^2}{2m}\frac{\partial^2}{\partial x^2} + V\right)}_{H_0} + \underbrace{\left(\frac{\hbar^2 k^2}{2m} + \frac{\hbar}{m}k\cdot\overbrace{\frac{\hbar}{i}\frac{\partial}{\partial x}}^{p}\right)}_{H' \text{ for small } k}\right] u = Eu \quad \left[\begin{array}{l}\text{a "Schrödinger eqn"} \\ \text{for the envelope fn} \\ u(x), \underline{\text{not }} \psi(x)\end{array}\right]$$

- Since $\frac{\hbar}{i}\frac{\partial}{\partial x} \equiv \hat{p}$, last term gives name "k-dot-p Theory"

- If we know the eigenenergies and eigenstates at $k=0$, we can use perturbation theory to approximate the true dispersion relation $E(k)$.

Two band k-dot-p dispersion

$$E_n(k) \cong E_n^0 + \langle \psi_n^0 | H' | \psi_n^0 \rangle + \sum_{\ell \ne n} \frac{|\langle \psi_n^0 | H' | \psi_\ell^0 \rangle|^2}{E_n - E_\ell}$$

$$H' = \frac{\hbar^2 k^2}{2m} + \frac{\hbar}{m}\vec{k}\cdot\vec{p}$$

$$E_c(k) = E_c + \frac{\hbar^2 k^2}{2m}\underbrace{\langle \psi_c^0 | \psi_c^0 \rangle}_{1 \text{ by normalization}} + \frac{\hbar}{m}\underbrace{\langle \psi_c^0 | \vec{k}\cdot\vec{p} | \psi_c^0 \rangle}_{0 \, @ \, k=0}$$

$$+ \underbrace{\frac{\hbar^2 k^2}{2m}\frac{|\langle \psi_c^0 | \psi_v^0 \rangle|^2}{E_c - E_v}}_{0 \text{ by orthogonality}} + \frac{\hbar^2}{m^2}\frac{|\langle \psi_c^0 | \vec{k}\cdot\vec{p} | \psi_v^0 \rangle|^2}{E_c - E_v}$$

So $$E_c(k) = E_c + \frac{\hbar^2 k^2}{2m} + \frac{\hbar^2 k^2}{m^2}\frac{|\langle \psi_c^0 | p | \psi_v^0 \rangle|^2}{E_g}$$

$$= E_c + \frac{\hbar^2 k^2}{2}\left(\frac{1}{m} + \frac{2P^2}{m^2 E_g}\right)$$

conduction: $$m^* = \left(\frac{1}{m} \pm \frac{2P^2}{m^2 E_g}\right)^{-1} \quad \text{"effective mass"}$$
(valence: $-$)

[Diagram: E(k) vs k showing parabolic conduction band with minimum E_c and valence band with maximum E_v, separated by bandgap E_g. "Conduction" states above, "valence" states below.]

[Plot: Effective mass (m_0) vs Bandgap (eV) showing approximately linear relationship; k·p theory with $2P^2/m_0 \sim 20$ eV; data points for InSb, InAs, Ge, GaSb, InP, GaAs, CdTe, GaN.]

Electronic "bandstructure" of crystalline solids

Physically, these periodic potentials are created by crystal lattices of atoms forming solids. In general, many more than two states are relevant, and the relative strength of nonzero off-diagonal matrix elements of $\hat{p}$ can result in unusual dispersion very different from the free-electron quadratic $\frac{\hbar^2 k^2}{2m^*}$.

For example, when gap-edge states are not coupled by $\hat{p}$, the valence band can have a $\sim k^4$ character, as in 2D

"Phosphorene"

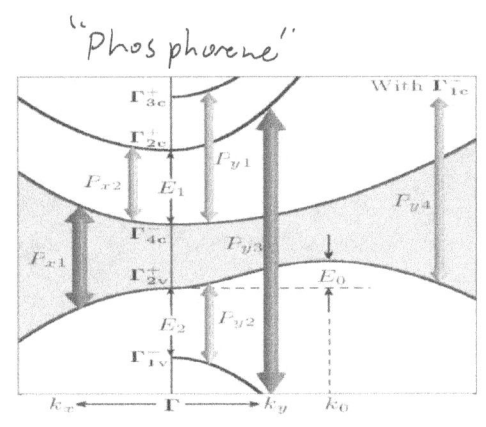

group-III chalcogenides (GaSe, etc)

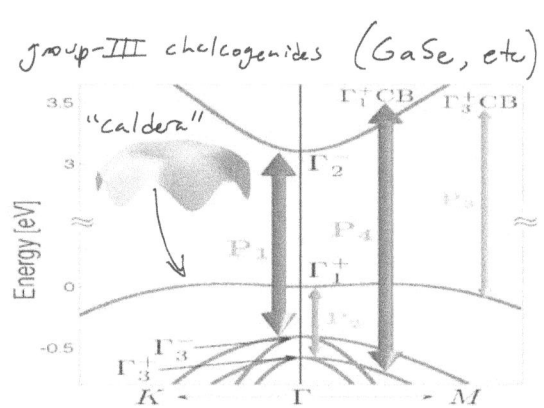

"caldera"

Potential distortion

- Infinite cubical well, $V=0$ inside

  ground state energy: $E_0 = \frac{\hbar^2}{2m}\left[\left(\frac{\pi}{a}\right)^2 + \left(\frac{\pi}{a}\right)^2 + \left(\frac{\pi}{a}\right)^2\right]$ $\xrightarrow{\text{distort}}$ $E_0 = \frac{\hbar^2}{2m}\left[\left(\frac{\pi}{a}\right)^2 + \left(\frac{\pi}{a}\right)^2 + \left(\frac{\pi}{b}\right)^2\right]$ (exact)

- What about a sphere?

  $E_1^0 = \frac{\pi^2 \hbar^2}{2m a^2}$ $\xrightarrow{\text{distort}}$ (ellipsoid, $b = (1+\epsilon)a$) $\quad V(x) = \begin{cases} 0 & \frac{x^2+y^2}{a^2} + \frac{z^2}{b^2} < 1 \\ \infty & \cdots \cdots \geq 1 \end{cases}$

- Restore broken spherical symmetry in (x,y,z') coordinates, $z = \frac{b}{a}z'$

  Then, T.I.S.E becomes $-\frac{\hbar^2}{2m}\left(\frac{\partial^2}{\partial x^2} + \frac{\partial^2}{\partial y^2} + \left(\frac{a}{b}\right)^2\frac{\partial^2}{\partial z'^2}\right)\psi = E\psi$, w/ b.c. $\psi(r'=a)=0$

- Expanding $\left(\frac{a}{b}\right)^2 = \frac{1}{(1+\epsilon)^2} \approx 1 - 2\epsilon$, we have a "perturbed" Hamiltonian

  $$\left\{\underbrace{-\frac{\hbar^2}{2m}\left(\frac{\partial^2}{\partial x^2} + \frac{\partial^2}{\partial y^2} + \frac{\partial^2}{\partial z^2}\right)}_{H^0} + \underbrace{\frac{\hbar^2}{2m}\cdot 2\epsilon \frac{\partial^2}{\partial z^2}}_{H'}\right\}\psi = E\psi$$

Ground state perturbation

Due to spherical symmetry of unperturbed Hamiltonian and $V_o(r)=0$,

$$\langle \psi_1^0 | \frac{\partial^2}{\partial z^2} | \psi_1^0 \rangle = \langle \psi_1^0 | \frac{\partial^2}{\partial x^2} | \psi_1^0 \rangle = \langle \psi_1^0 | \frac{\partial^2}{\partial y^2} | \psi_1^0 \rangle = \frac{1}{3}\left(-\frac{2m}{\hbar^2} E_1^0\right)$$

So first-order perturbation correction is

$$\langle \psi_1^0 | H' | \psi_1^0 \rangle = -\frac{2\epsilon}{3} E_1^0 \quad \text{and} \quad E_1 \approx \left(1 - \frac{2}{3}\epsilon\right) \frac{\pi^2 \hbar^2}{2m a^2}$$

- Note that ellipsoid volume is
$$V = \frac{4}{3}\pi a^2 b = \frac{4}{3}\pi a^3 (1+\epsilon) \equiv \frac{4}{3}\pi R^3, \quad \text{where } R = a(1+\epsilon)^{1/3} \approx a\left(1+\frac{\epsilon}{3}\right)$$
is radius of a sphere w/ same volume as our ellipsoid.

The ground state energy of this sphere is
$$E_1 = \frac{\pi^2 \hbar^2}{2m R^2} = \frac{\pi^2 \hbar^2}{2m a^2}\left(1+\frac{\epsilon}{3}\right)^{-2} \approx \left(1-\frac{2}{3}\epsilon\right)\frac{\pi^2 \hbar^2}{2m a^2} \quad \text{c.f. above!}$$

- So all <u>volume-preserving</u> deformations have vanishing lowest-order corrections to ground-state energy!

Approximating the ground state energy

Without solving $H|\psi_n\rangle = E|\psi_n\rangle$ for all n, we can <u>guess</u> a ground-state wave function. Since the <u>true</u> eigenfunctions of H form a complete orthonormal set, conceptually we have $|\psi\rangle = \sum_n c_n |\psi_n\rangle$, with normalization $\sum_n |c_n|^2 = 1$.

Therefore $\langle \psi | H | \psi \rangle = \langle \psi | c_1 H | \psi_1 \rangle + \langle \psi | c_2 H | \psi_2 \rangle + \cdots$

$$= c_1 E_1 \langle \psi | \psi_1 \rangle + c_2 E_2 \langle \psi | \psi_2 \rangle + \cdots$$

By orthonormality: $= c_1 E_1 (\langle \psi_1 | c_1^* + \cancel{\langle \psi_2 | c_2^*}^{0} + \cdots) |\psi_1\rangle + c_2 E_2 (\cancel{\langle \psi_1 | c_1^*}^{0} + \langle \psi_2 | c_2^* + \cdots) |\psi_2\rangle + \cdots$

$$= |c_1|^2 E_1 (+ |c_2|^2 E_2 + \cdots) = |c_1|^2 E_1 (+|c_2|^2 (E_1 + \Delta_2) + \cdots)$$

$$= E_1 \sum_n |c_n|^2 + \sum_{n \neq 1} |c_n|^2 \Delta_n$$

Since $|c_n|^2 \geq 0$ and $E_n > E_1$ for $n>1$ ($\Delta_n > 0$), $\boxed{\langle \psi | H | \psi \rangle \geq E_1}$

So we can place an upper bound on the true ground state energy!

Upper bound on first excited state energy

Under certain circumstances we can even use this method to put an upper bound on the <u>first excited state</u> energy, by exploiting a known symmetry of the ground state. Choosing a $|\varphi\rangle$ orthogonal to the unknown $|\psi_1\rangle$ such that $\langle\varphi|\psi_1\rangle = 0$ <u>by symmetry</u> eliminates the first term in

$$\langle\varphi|H|\varphi\rangle = c_1 E_1 \langle\varphi|\psi_1\rangle + c_2 E_2 \langle\varphi|\psi_2\rangle + \ldots$$

and we can then show $\langle\varphi|H|\varphi\rangle \geq E_2$.

Example: If H is spatially even, the ground state wave function is even. So, choose an <u>odd</u> function as a guess. We know this choice is orthogonal to the ground state, despite not knowing what the ground state wavefunction is!

Example: Harmonic oscillator

$H = -\frac{\hbar^2}{2m}\frac{d^2}{dx^2} + \frac{1}{2}m\omega^2 x^2$. "Educated" guess: $\varphi = Ae^{-bx^2}$, calculate $\langle\varphi|H|\varphi\rangle$

First, normalize φ:

$$\int_{-\infty}^{\infty} \varphi^*\varphi\, dx = \int_{-\infty}^{\infty} A^* e^{-bx^2} \cdot A e^{-bx^2} dx = |A|^2 \int_{-\infty}^{\infty} e^{-2bx^2} dx = |A|^2 \sqrt{\frac{\pi}{2b}} = 1. \quad \text{So } |A|^2 = \sqrt{\frac{2b}{\pi}}.$$

Now,

$$\langle\varphi|H|\varphi\rangle = \int_{-\infty}^{\infty} A^* e^{-bx^2}\left(-\frac{\hbar^2}{2m}\frac{d^2}{dx^2} + \frac{1}{2}m\omega^2 x^2\right) A e^{-bx^2} dx$$

$$= |A|^2 \left[\frac{1}{2m}\int_{-\infty}^{\infty} e^{-bx^2}\left(\frac{\hbar}{i}\frac{d}{dx}\right)^2 e^{-bx^2} dx + \frac{1}{2}m\omega^2 \int_{-\infty}^{\infty} e^{-2bx^2} x^2 dx\right]$$

Because $\hat{p} = \frac{\hbar}{i}\frac{d}{dx}$ is hermitian, $\langle\varphi|\hat{p}^2|\varphi\rangle = |\hat{p}|\varphi\rangle|^2$ and the first term is:

$$\frac{1}{2m}\int_{-\infty}^{\infty} \frac{\hbar}{i} e^{-bx^2}(-2bx) \cdot \frac{\hbar}{i} e^{-bx^2}(-2bx)\, dx = \frac{4b^2\hbar^2}{2m}\int_{-\infty}^{\infty} e^{-2bx^2} x^2 dx$$

Variational minimization

To evaluate, use derivative under integral sign:

$$\int_{-\infty}^{\infty} e^{-2bx^2} x^2 dx = \int_{-\infty}^{\infty}\left(-\frac{1}{2}\frac{d}{db}\right)e^{-2bx^2} dx = -\frac{1}{2}\frac{d}{db}\int_{-\infty}^{\infty} e^{-2bx^2} dx = -\frac{1}{2}\frac{d}{db}\sqrt{\frac{\pi}{2b}} = \frac{\sqrt{\pi}}{4}\frac{b^{-3/2}}{\sqrt{2}}$$

Then,

$$\langle \psi | H | \psi \rangle = \sqrt{\frac{2b}{\pi}}\left[\frac{\hbar^2 b^2}{2m} \cdot \frac{\sqrt{\pi}}{4}\frac{b^{-3/2}}{\sqrt{2}} + \frac{1}{2}m\omega^2 \cdot \frac{\sqrt{\pi}}{4}\frac{b^{-3/2}}{\sqrt{2}}\right] = \frac{\hbar^2 b}{2m} + \frac{m\omega^2}{8b}.$$

The best possible upper bound to g.s. energy is the _lowest_ possible value of expression above:

$$\frac{d}{db}\left(\frac{\hbar^2 b}{2m} + \frac{m\omega^2}{8b}\right) = \frac{\hbar^2}{2m} - \frac{m\omega^2}{8b^2} = 0 \Rightarrow \frac{\hbar^2}{2m} = \frac{m\omega^2}{8b^2} \Rightarrow b^2 = \frac{m\omega^2}{\frac{8\hbar^2}{2m}} = \frac{m^2\omega^2}{4\hbar^2}$$

So, optimum value is $b = \frac{m\omega}{2\hbar}$, which has correct units: $\frac{g \cdot \frac{1}{s}}{erg \cdot s} = \frac{1}{cm^2}$ ✓

For this b, $\langle \psi | H | \psi \rangle = \frac{\hbar^2}{2m}\frac{m\omega}{2\hbar} + \frac{m\omega^2}{8}\frac{2\hbar}{m\omega} = \frac{\hbar\omega}{4} + \frac{\hbar\omega}{4} = \frac{\hbar\omega}{2}$

Same as the _true_ ground state energy, because we chose ψ "wisely"!

A not-so-wise choice: gaussian for the delta potential

$H = -\frac{\hbar^2}{2m}\frac{d^2}{dx^2} - \alpha\delta(x)$ Guess: $\psi(x) = \left(\frac{2b}{\pi}\right)^{1/4} e^{-bx^2}$ (Not the true g.s. wavefunction)

$$\langle \psi | H | \psi \rangle = \frac{\hbar^2 b}{2m} - \alpha\left(\frac{2b}{\pi}\right)^{1/2}\int_{-\infty}^{\infty} e^{-2bx^2}\delta(x) dx = \frac{\hbar^2 b}{2m} - \alpha\left(\frac{2b}{\pi}\right)^{1/2}$$

↖ Kinetic energy expectation value, from previous calculation.

Now minimize:

$$\frac{d}{db}\langle \psi | H | \psi \rangle = \frac{\hbar^2}{2m} - \frac{\alpha}{\sqrt{2b\pi}} = 0 \Rightarrow b = \frac{\alpha^2 4m^2}{2\pi\hbar^4} = \frac{2m^2\alpha^2}{\pi\hbar^4}$$

So, our best guess for the ground-state energy w/ this wavefunction is

$$E_1 \leq \frac{\hbar^2}{2m}\frac{2m^2\alpha^2}{\pi\hbar^4} - \alpha\left(\frac{2 \cdot 2m^2\alpha^2}{\pi^2\hbar^4}\right)^{1/2} = \frac{m\alpha^2}{\pi\hbar^2} - \frac{2m\alpha^2}{\pi\hbar^2} = -\frac{m\alpha^2}{\pi\hbar^2} \left(\frac{erg \cdot g \cdot erg \cdot cm^2}{(erg \cdot s)^2} = eV\ ✓\right)$$

Not bad: it is certainly bound ($E<0$) and (knowing the exact eigenvalue) it has the correct dependence on α. Can we do better?

A better choice of variational wavefunction

Now try $\psi(x) = Ae^{-b|x|}$ for $V(x) = -\alpha\delta(x)$:

Normalize: $\int_{-\infty}^{\infty} A^* e^{-b|x|} A e^{-b|x|} dx = 2|A|^2 \int_0^{\infty} e^{-2bx} dx = 2|A|^2 \left.\frac{e^{-2bx}}{-2b}\right|_0^{\infty} = \frac{|A|^2}{b} = 1$

So $A = \sqrt{b}$ and $\langle\psi|H|\psi\rangle = -\frac{\hbar^2}{2m}\int_{-\infty}^{\infty}\sqrt{b}\,e^{-b|x|}\left(\frac{d}{dx}\right)^2\sqrt{b}\,e^{-b|x|}dx - \alpha\int_{-\infty}^{\infty} b\cdot e^{-2b|x|}\delta(x)dx$

We need to be careful w/ the first integral because the wavefunction has a 1st derivative discontinuity @ $x=0$, where $\psi'' = -2b^{3/2}\delta(x)$:

$-\frac{\hbar^2}{2m}\left(2\int_0^{\infty} b^3 e^{-2bx}dx + \lim_{\epsilon\to 0}\int_{-\epsilon}^{\epsilon}\sqrt{b}\,e^{-b|x|}(-2b^{3/2}\delta(x))dx\right) - \alpha b = -\frac{\hbar^2}{2m}(b^2 - 2b^2) - \alpha b = \frac{\hbar^2 b^2}{2m} - \alpha b$

Minimize: $\frac{d}{db}\left(\frac{\hbar^2}{2m}b^2 - \alpha b\right) = \frac{\hbar^2 b}{m} - \alpha = 0 \Rightarrow b = \frac{\alpha m}{\hbar^2}$, so: $E_1 \leq \frac{\hbar^2}{2m}\frac{\alpha^2 m^2}{\hbar^4} - \alpha\frac{\alpha m}{\hbar^2} = -\frac{\alpha^2 m}{2\hbar^2}$

This is lower than our previous best-guess, because we used the true (exact) ground-state wavefunction!

Two electrostatically interacting electrons: neutral Helium

The power of the variational method is in applying it to problems for which no exact solution exists! Take the two-electron Hamiltonian:

$$H = -\frac{\hbar^2}{2m}(\nabla_1^2 + \nabla_2^2) - \frac{e^2}{4\pi\epsilon_0}\left(\frac{2}{r_1} + \frac{2}{r_2} - \frac{1}{|\vec{r}_1 - \vec{r}_2|}\right)$$

The interaction term (Coulomb repulsion) depending on both electron coordinates makes this problem impossible to solve analytically. Our choice of variational wavefunction is motivated by our intuition that, due to electrostatic "screening" by the other electron, each behaves as it would in a single-electron atom, but with less than $+2$ nuclear charge:

$$\psi(\vec{r}_1, \vec{r}_2) = \psi_1(r_1)\psi_2(r_2) = \sqrt{\frac{Z^3}{\pi a^3}}\,e^{-Zr_1/a} \cdot \sqrt{\frac{Z^3}{\pi a^3}}\,e^{-Zr_2/a} = \frac{Z^3}{\pi a^3}e^{-Z(r_1+r_2)/a} \quad (a \text{ is Bohr radius})$$

where we expect $Z < 2$ minimizes $\langle\psi|H|\psi\rangle$. If we add + subtract electron-nuclear potential energy for nucleus w/ charge $+Ze$,

$$H = -\frac{\hbar^2}{2m}(\nabla_1^2 + \nabla_2^2) - \frac{e^2}{4\pi\epsilon_0}\left(\frac{Z}{r_1} + \frac{Z}{r_2}\right) + \frac{e^2}{4\pi\epsilon_0}\left(\frac{Z-2}{r_1} + \frac{Z-2}{r_2} + \frac{1}{|\vec{r}_1-\vec{r}_2|}\right)$$

Then, the expectation value of the first two terms are just twice Bohr's result.

Single-particle expectation value <1/r>

$$\langle \psi | H | \psi \rangle = 2 \times \left(Z^2 \overbrace{E_1}^{-13.6 eV} + \frac{e^2}{4\pi\epsilon_0}(Z-2)\langle \tfrac{1}{r}\rangle \right) + \langle \psi | \frac{e^2}{4\pi\epsilon_0 |\vec{r}_1 - \vec{r}_2|} | \psi \rangle$$

Now we need to evaluate the remaining two terms. First, the term that only depends on electron coordinates one at a time:

$$\langle \tfrac{1}{r}\rangle = \int_0^{2\pi}\int_0^{\pi}\int_0^{\infty} \left(\frac{Z^3}{\pi a^3}\right)^{1/2} e^{-\frac{Zr}{a}} \frac{1}{r} \left(\frac{Z^3}{\pi a^3}\right)^{1/2} e^{-Zr/a} \left(r^2 dr\, \sin\theta\, d\theta\, d\phi\right)$$

Since integrand does not depend on θ or ϕ,

$$= 4\pi \frac{Z^3}{\pi a^3} \int_0^{\infty} e^{-2Zr/a}\, r\, dr = \frac{4Z^3}{a^3}\left(-\frac{a}{2}\frac{d}{dZ}\int_0^{\infty} e^{-2Zr/a}\, dr\right)$$

$$= \frac{4Z^3}{a^3}\left(-\frac{a}{2}\frac{d}{dZ}\left(-\frac{a}{2Z}e^{-2Zr/a}\Big|_0^{\infty}\right)\right) = -\frac{Z^3}{a}\frac{d}{dZ}\frac{1}{Z} = \frac{Z^3}{a}\frac{1}{Z^2} = \frac{Z}{a} \left(=\frac{1}{\langle r \rangle}\right)$$

This has the right units and makes sense physically.

Coulomb repulsion term

$$\langle \psi | \frac{1}{|\vec{r}_1 - \vec{r}_2|} | \psi \rangle = \iint \psi^* \frac{1}{|\vec{r}_1 - \vec{r}_2|} \psi\, d^3r_1\, d^3r_2$$

For $\vec{r}_2$ integral, align $\vec{r}_1$ to polar axis. Then, $|\vec{r}_1 - \vec{r}_2| = \sqrt{r_1^2 + r_2^2 - 2r_1 r_2 \cos\theta_2}$

$$\langle \psi | \frac{1}{|\vec{r}_1-\vec{r}_2|} | \psi \rangle = \left(\frac{Z^3}{\pi a^3}\right)^2 \int e^{-2Zr_1/a} \boxed{\int \frac{e^{-2Zr_2/a}}{\sqrt{r_1^2+r_2^2-2r_1 r_2 \cos\theta_2}} r_2^2 \sin\theta_2\, d\theta_2\, d\phi_2\, dr_2}\, r_1^2 \sin\theta_1\, d\theta_1\, d\phi_1\, dr_1$$

$$\boxed{} = 2\pi \int_0^{\infty} r_2^2\, e^{-2Zr_2/a}\, \frac{\sqrt{r_1^2+r_2^2-2r_1 r_2 \cos\theta}}{r_1 r_2}\Bigg|_{\theta_2=0}^{\pi} dr_2$$

$$= 2\pi \int_0^{\infty} e^{-2Zr_2/a} \frac{r_2}{r_1}\boxed{\left(\sqrt{r_1^2+r_2^2+2r_1 r_2} - \sqrt{r_1^2+r_2^2-2r_1 r_2}\right)} dr_2$$

$$\boxed{} = r_1 + r_2 - |r_1 - r_2| = \begin{cases} 2r_2 & \text{if } r_2 < r_1 \\ 2r_1 & \text{if } r_2 > r_1 \end{cases}$$

Splitting the r₂ integral

$$\square = 4\pi\left[\int_0^{r_1} e^{-2Zr_2/a}\frac{r_2^2}{r_1}dr_2 + \int_{r_1}^{\infty} e^{-2Zr_2/a} r_2\, dr_2\right]$$

$(r_2 < r_1)$ $(r_2 > r_1)$

Using differentiation under integral sign (twice for the first term)

$$= \frac{\pi a^3}{Z^3 r_1}\left(1 - \left(1 + \frac{Zr_1}{a}\right)e^{-2Zr_1/a}\right)$$

Inserting into our expectation value:

$$\langle \psi | \frac{1}{|\vec{r}_1 - \vec{r}_2|} | \psi \rangle = \left(\frac{Z^3}{\pi a^3}\right)^2 \int e^{-2Zr_1/a}\boxed{\left(\frac{\pi a^3}{Z^3 r_1}\right)\left(1-\left(1+\frac{Zr_1}{a}\right)e^{-2Zr_1/a}\right)} r_1^2 \sin\theta_1\, d\theta_1\, d\phi_1\, dr_1$$

$$= \frac{Z^3}{\pi a^3} \int e^{-2Zr_1/a}\left(1-\left(1+\frac{Zr_1}{a}\right)e^{-2Zr_1/a}\right) r_1 \underbrace{\sin\theta_1\, d\theta_1\, d\phi_1}_{\to 4\pi}\, dr_1$$

$$= \frac{4Z^3}{a^3}\int_0^{\infty} e^{-2Zr_1/a}\left(1-\left(1+\frac{Zr_1}{a}\right)e^{-2Zr_1/a}\right) r_1\, dr_1$$

Evaluating expectation value

$$\langle \frac{1}{|\vec{r}_1 - \vec{r}_2|}\rangle = \frac{4Z^3}{a^3}\left[\int_0^{\infty} e^{-2Zr/a} r\, dr - \int_0^{\infty} e^{-4Zr/a} r\, dr - \frac{Z}{a}\int_0^{\infty} e^{-4Zr/a} r^2 dr\right]$$

Differentiating under the integral sign several times gives:

$$= \frac{4Z^3}{a^3}\left[-\frac{a}{2}\frac{d}{dZ}\int_0^{\infty} e^{-2Zr/a}dr + \frac{a}{4}\frac{d}{dZ}\int_0^{\infty} e^{-4Zr/a}dr - \frac{Z}{a}\left(\frac{a}{4}\right)^2\frac{d^2}{dZ^2}\int_0^{\infty} e^{-4Zr/a}dr\right]$$

$$= \frac{4Z^3}{a^3}\left[+\frac{a^2}{4Z^2} - \frac{a^2}{16Z^2} + \frac{Za}{16}\frac{d}{dZ}\left(\frac{a}{4Z^2}\right)\right]$$

$$= \frac{Z}{a}\left[1 - \frac{1}{4} - \frac{1}{8}\right] = \frac{5Z}{8a}$$

So finally we have

$$\langle \psi | H | \psi \rangle = 2\left(Z^2 E_1 + \frac{e^2}{4\pi\varepsilon_0}(Z-2)\frac{Z}{a}\right) + \frac{e^2}{4\pi\varepsilon_0}\left(\frac{5Z}{8a}\right)$$

Variational minimization

Now, note that $\frac{e^2}{4\pi\epsilon_0 a} = \overset{\left(\frac{e^2}{4\pi\epsilon_0}\right)}{\alpha \hbar c} \cdot \overset{\left(\frac{1}{a}\right)}{\frac{mc\alpha}{\hbar}} = \alpha^2 mc^2 = -2\left(-\frac{1}{2}\alpha^2 mc^2\right) = -2E_1$.

Thus, $\langle \psi|H|\psi\rangle = 2Z^2 E_1 - 4Z(Z-2)E_1 - \frac{5Z}{4}E_1 = \left(-2Z^2 + \frac{27}{4}Z\right)E_1$

Minimizing,

$$\frac{d}{dZ}\left(-2Z^2 + \frac{27}{4}Z\right) = -4Z + \frac{27}{4} = 0 \implies Z = \frac{27}{16} = \frac{1}{2}\left(\frac{3}{2}\right)^3 \sim 1.7$$

This result is consistent with our intuition that motivated our choice of variational wavefunction since $Z < 2$ minimizes the energy:

$$\langle \psi|H|\psi\rangle = \left(-\frac{1}{2}\left(\frac{3}{2}\right)^6 + 2\left(\frac{3}{2}\right)^3 \frac{1}{2}\left(\frac{3}{2}\right)^3\right)E_1 = \frac{1}{2}\left(\frac{3}{2}\right)^6 E_1 \sim -77.5 \, eV$$

Compare to experimental value of $-79.975 \, eV$!

Semiclassical/WKB approximation   "Wentzel – Kramers – Brillouin"

- Classically $E = \frac{p^2}{2m} + V$, so we can write $p = \sqrt{2m(E-V)}$. Extending to corresponding quantum mechanical operators, the 1-d Schrödinger eqn:
$\left[-\frac{\hbar^2}{2m}\frac{d^2}{dx^2} + V(x)\right]\psi(x) = E\psi(x)$ can be written as $\frac{d^2\psi}{dx^2} = -\frac{p^2}{\hbar^2}\psi$.

- The unknown complex scalar-valued wavefunction $\psi(x)$ can be expressed, without any loss of generality, in "polar" form $\psi(x) = A(x)e^{i\phi(x)}$, where $A(x)$ and $\phi(x)$ are <u>real-valued</u> functions.

- Then, two applications of the derivative product rule $(fg)' = fg' + gf'$ leaves us with four terms on the LHS, all proportional to $e^{i\phi}$:

$$\left(A'' + 2iA'\phi' + iA\phi'' - A(\phi')^2\right)e^{i\phi} = -\frac{p^2}{\hbar^2}A e^{i\phi}$$

Note that this can be written as two (coupled) diff. eqs: for real and imaginary parts!

Real and imaginary parts

Collecting and equating real and imaginary parts gives two equations:

real part: $A'' - A(\phi')^2 = -\frac{p^2}{\hbar^2} A$

imaginary part: $2iA'\phi' + iA\phi'' = 0$

• Imaginary part:

Note that $i(A^2\phi')' = A(2iA'\phi' + iA\phi'')$. Then we can solve $(A^2\phi')' = 0 \implies A(x) = \frac{C}{\sqrt{\phi'(x)}}$, where C is an undetermined const. of integration.

• Real part: Here is where the approximation occurs. When the magnitude of the wavefunction $A(x)$ varies slowly, A'' is negligible. Then we have a solvable differential equation!

$$-A(\phi')^2 = -\frac{p^2}{\hbar^2} A \implies \phi(x) = \pm\frac{1}{\hbar}\int_0^x p(x')\, dx'$$

Thus, our WKB wavefunction is

$$\psi(x) = A(x) e^{i\phi(x)} \cong \frac{C}{\sqrt{p(x)}} e^{\pm\frac{i}{\hbar}\int_0^x p(x')\,dx'}$$

Note that this is invalid for $p(x) \approx 0$, i.e. @ "classical turning point".

WKB perspective on tunneling thru single barrier

If $E \ll V(x)$,

$$\psi^{WKB}(x) = \frac{C_+}{\sqrt{|p(x)|}} e^{|\phi(x)|} + \frac{C_-}{\sqrt{|p(x)|}} e^{-|\phi(x)|}$$

For energy far below the top of the barrier, exponential decay of the evanescent wavefunction dominates, so little probability flux reflected from right interface makes it back to the left interface to cause interference. $\to C_+ \approx 0$

We can therefore approximate total transmission by the product of three incoherent events:

1. transmission thru left interface
2. exponential decay thru barrier: $T_{WKB} = \left|\frac{\psi(d)}{\psi(0)}\right|^2 = e^{-2\gamma}$, $\gamma = \frac{1}{\hbar}\int_0^d |p(x)|\, dx$
3. transmission thru right interface.

Step potential

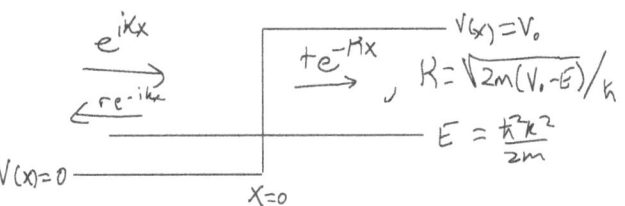

our left interface looks like a step,

- Boundary conditions

  $\psi(x)$ continuous: $\left(e^{ikx} + re^{-ikx} = te^{-\kappa x}\right)\Big|_{x=0} \rightarrow 1+r=t$

  $\psi'(x)$ continuous: $\left(ike^{ikx} - ikre^{-ikx} = -\kappa te^{-\kappa x}\right)\Big|_{x=0} \rightarrow ik(1-r) = -\kappa t$

  $t = \frac{ik}{\kappa}(r-1) = \frac{ik}{\kappa}(t-2) \rightarrow t\left(1 - \frac{ik}{\kappa}\right) = \frac{-2ik}{\kappa} \rightarrow t = \frac{-2ik}{\kappa - ik}$

- $T_{left} = \frac{J_{trans}}{J_{inc}} = \frac{\frac{\hbar}{m}\text{Im}(t^* \cdot (-\kappa)t)}{\frac{\hbar}{m}\text{Im}(ik)} = \frac{-\kappa}{ik}|t|^2 = \frac{\kappa i}{k}\left(\frac{2ik}{\kappa+ik}\right)\left(\frac{-2ik}{\kappa-ik}\right) = \frac{4k\kappa i}{k^2+\kappa^2}$

  $\left[J = \frac{\hbar}{m}\text{Im}(\psi^* \nabla \psi)\right]$

- At right interface, $(ik \mid \kappa) \rightarrow (\kappa \mid ik)$, $T_{right} = \frac{-4k\kappa i}{k^2+\kappa^2}$

- $T = T_{left}\, T_{WKB}\, T_{right} = \left(\frac{4k\kappa}{k^2+\kappa^2}\right)^2 e^{-2\gamma}$, $\gamma = \frac{1}{\hbar}\int_0^d \sqrt{2m(V(x)-E)}\, dx$

Increasing barrier width

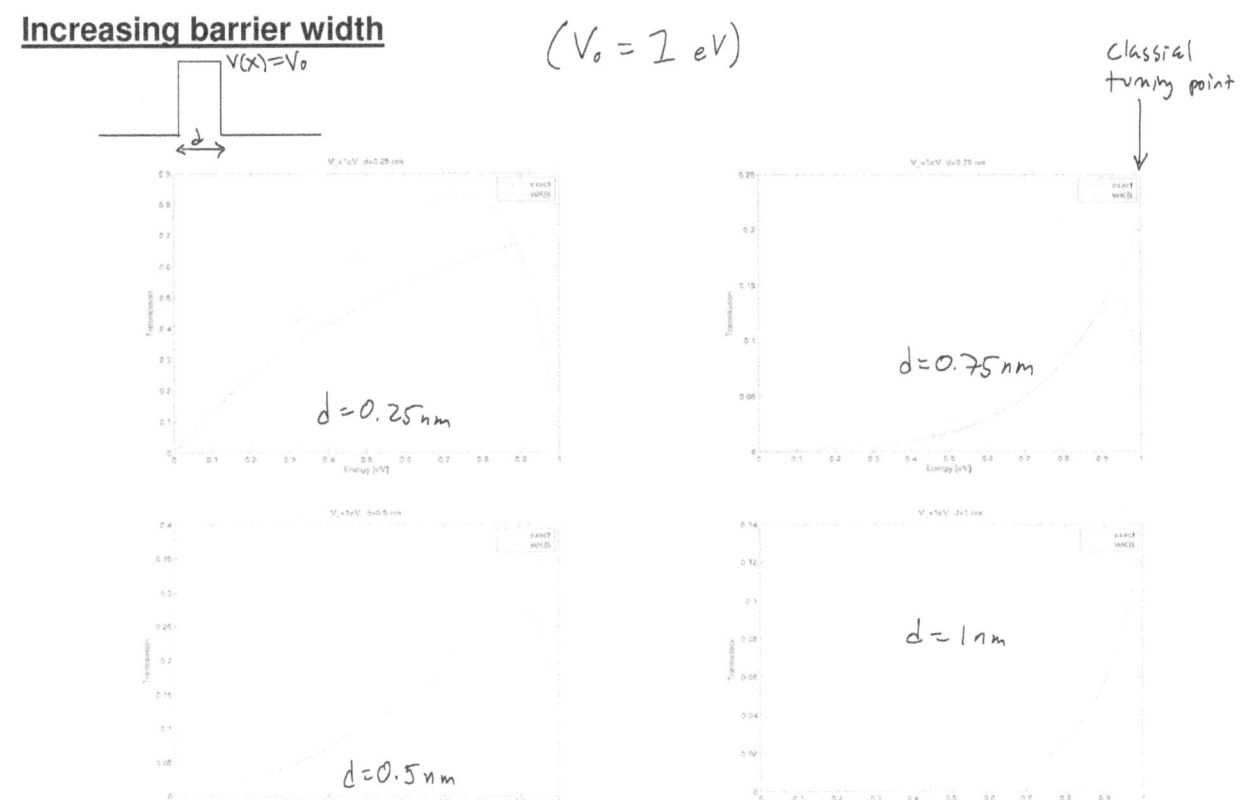

$(V_0 = 1\, eV)$

classical turning point

Increasing barrier height $(d = 0.25 \text{ nm})$

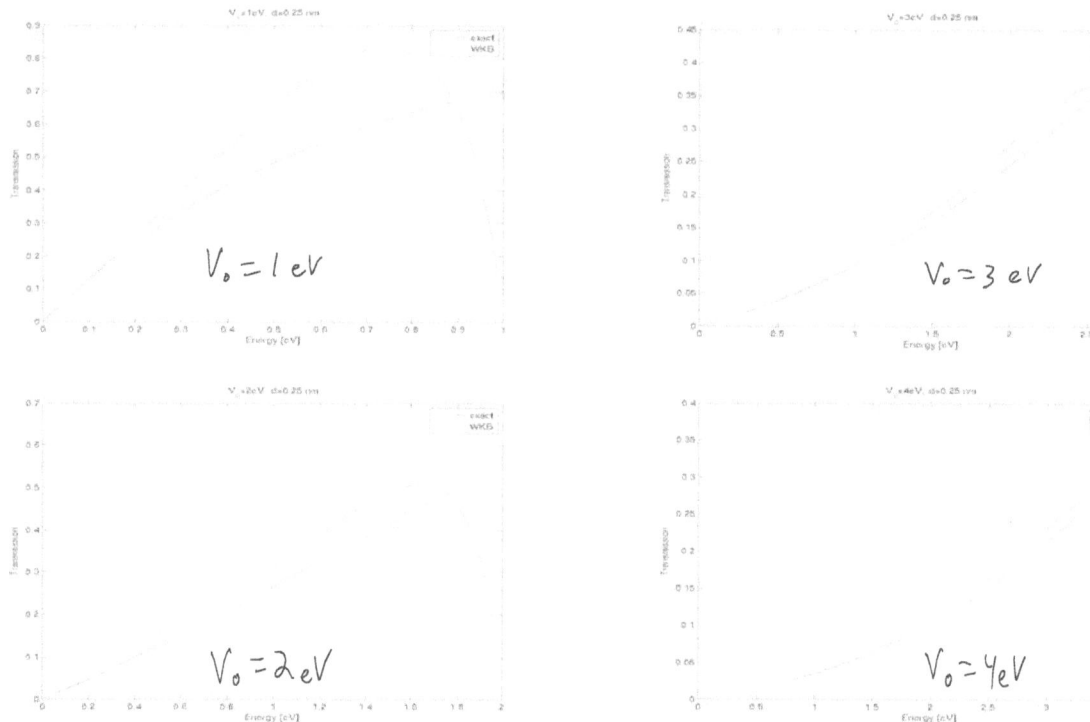

"Tunnel junction"

In a "tunnel junction", two metals are separated by an insulator. A net flow of electrons from the "Fermi sea" of one metal to the other is induced by creating an asymmetry of filled/empty states w/ voltage drop:

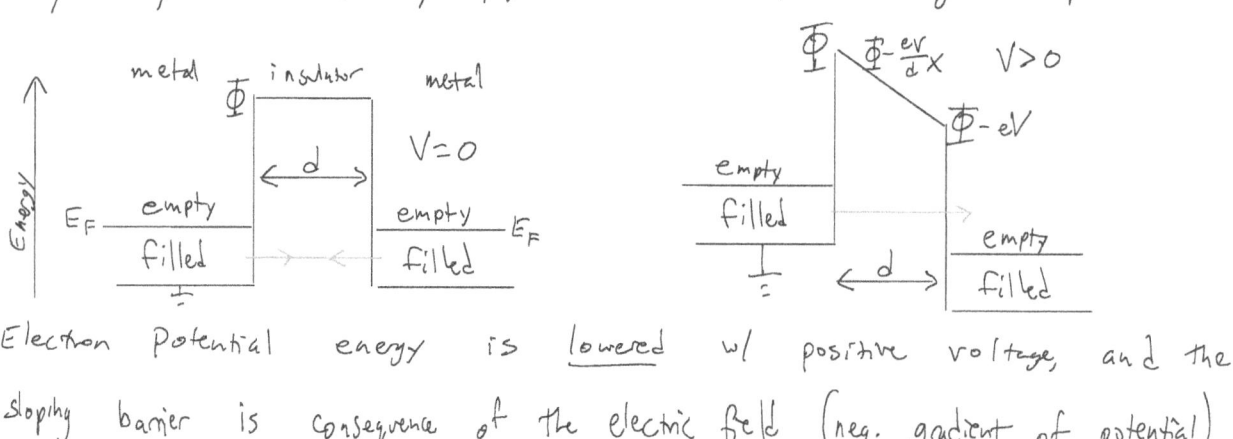

Electron potential energy is <u>lowered</u> w/ positive voltage, and the sloping barrier is consequence of the electric field (neg. gradient of potential).

Transmission coef. of trapezoidal barrier

Assuming $eV \ll \Phi - E_F$ consistent w/ WKB constraints, the dominant component in WKB transmission coefficient is

$$T_{WKB} \cong e^{-2\gamma}, \quad \gamma = \frac{1}{\hbar}\int_{x_1}^{x_2} |p(x)|\, dx,$$

where x_1 and x_2 are the classical turning points. Here,

$$\gamma = \frac{1}{\hbar}\int_0^d \sqrt{2m\left(\Phi - \frac{eV}{d}x - E\right)}\, dx$$

$$= -\frac{2}{3}\frac{\sqrt{2m}}{\hbar}\left(\Phi - \frac{eV}{d}x - E\right)^{3/2} \frac{d}{eV}\bigg|_0^d$$

$$= \frac{2}{3}\frac{\sqrt{2m}}{\hbar}\frac{d}{eV}\left((\Phi - E)^{3/2} - (\Phi - eV - E)^{3/2}\right)$$

$$= \frac{2}{3}\frac{\sqrt{2m}}{\hbar}\frac{d}{eV}\left((\Phi - E)^{3/2} - (\Phi - E)^{3/2}\left(1 - \frac{eV}{\Phi - E}\right)^{3/2}\right)$$

An insightful/useful approximation

To second order, $(1+x)^n \approx 1 + nx + \frac{n(n-1)}{2}x^2$, so again assuming $\frac{eV}{\Phi-E} \ll 1$,

$$\gamma \approx \frac{2}{3}\frac{\sqrt{2m}}{\hbar}\frac{d}{eV}\left((\Phi - E)^{3/2} - (\Phi - E)^{3/2}\left(\cancel{1} - \frac{3}{2}\frac{eV}{\Phi - E} + \frac{3}{2}\frac{1}{4}\left(\frac{eV}{\Phi - E}\right)^2\right)\right)$$

$$= \frac{\sqrt{2m}}{\hbar}d\left((\Phi - E)^{1/2} - \frac{1}{4}\frac{eV}{(\Phi - E)^{1/2}}\right) = \underbrace{\frac{\sqrt{2m}}{\hbar}d(\Phi - E)^{1/2}}_{\gamma(V=0)}\underbrace{\left(1 - \frac{eV}{4(\Phi - E)}\right)}_{V \neq 0 \text{ correction}}$$

So nonzero voltage lowers γ and results in higher $T \cong e^{-2\gamma}$. We can gain some insight by approximating the correction w/ (inverse) Taylor expansion:

$$1 - \frac{eV}{4(\Phi - E)} \approx \left(1 - \frac{eV}{2(\Phi - E)}\right)^{1/2}$$

giving:

$$\gamma \approx \frac{\sqrt{2m}\, d}{\hbar}\sqrt{\left(\Phi - \frac{eV}{2}\right) - E} = \frac{1}{\hbar}\int_0^d \sqrt{2m\left(\Phi - \frac{eV}{2} - E\right)}\, dx$$

So trapezoidal distortion is equivalent to barrier lowering by an amount equal to half the potential drop, to the _average height_!

Numerical comparison between methods

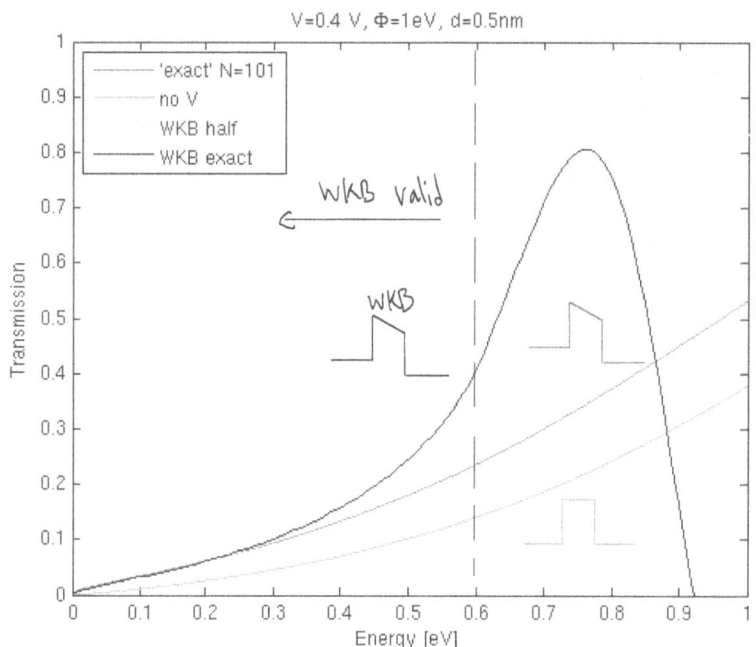

Bound-state WKB (semiclassical) approximation

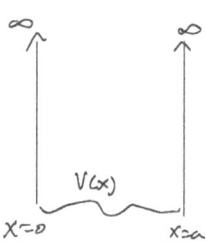

When $E \gg V(x)$,

$$\psi_{WKB}(x) = \frac{C_+}{\sqrt{p(x)}} e^{+i\phi(x)} + \frac{C_-}{\sqrt{p(x)}} e^{-i\phi(x)}$$

Assume a potential $V(x)$ w/ infinite barriers at $x=0,a$:

It is convenient to take even+odd superpositions of $e^{\pm i\phi(x)}$ fns:

$$\psi_{WKB}(x) = \frac{1}{\sqrt{p(x)}} \left[C_1 \sin\phi(x) + C_2 \cos\phi(x) \right]$$

Boundary conditions $(\psi(0) = \psi(a) = 0)$ imply $C_2 = 0$, zeros of $\sin\phi$ give

$$\phi(a) = \frac{1}{\hbar} \int_0^a \sqrt{2m(E-V(x))}\, dx = n\pi, \quad n=1,2,3\ldots$$

Trivial example: $V(x)=0$, $0<x<a$ as in the "infinite QW". Then

$$\phi(a) = \frac{1}{\hbar} \int_0^a \sqrt{2mE}\, dx = \frac{\sqrt{2mE}\,a}{\hbar} = n\pi \implies E = \frac{(n\pi\hbar)^2}{2ma^2}$$

Our approx is exact since the WKB wavefunction ampl. is constant in $0<x<a$

Beyond the quantum well

The above approach only works because $\psi(x)=0$ when $V\to\infty > E$, so we can avoid the divergence of $1/\sqrt{p(x)}$ term in WKB wavefunction. What happens when the potential is finite?

Then $\psi(x)$ is nonzero at the classical turning pts around which WKB is invalid!

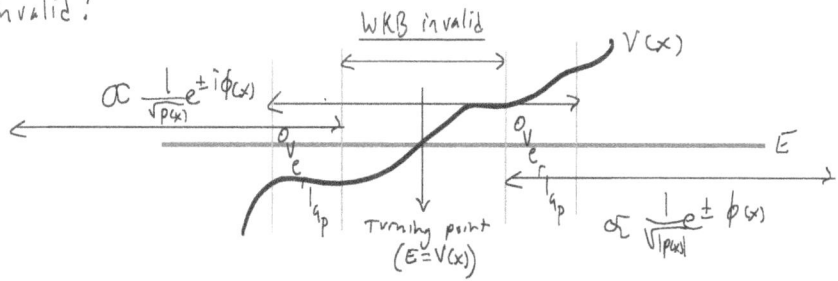

Solution: solve the Schrodinger equation exactly around the turning point and match it to the WKB functions at edges of "overlap regions" far enough away that WKB is valid.

At the turning point

Assume that $V(x)$ is linear around turning pt: $V(x) \approx E + \frac{dV}{dx}\big|_{x=x_t} x$

Our Schrodinger eqn becomes: $-\frac{\hbar^2}{2m}\psi'' + (E + V'x)\psi = E\psi$ (No longer an eigenvalue eq!)

This is equivalent to $\frac{d^2\psi}{dz^2} = z\psi$, $z = \alpha x$, $\alpha = \left[\frac{2m}{\hbar^2} V'\right]^{1/3}$

Solutions are "Airy functions" $Ai(z)$, $Bi(z)$

Matching the asymptotic forms of these functions to the WKB wavefcns results in the following simple result:

$$\psi_{WKB}(x) = \frac{C}{\sqrt{p(x)}} \sin\left(\frac{1}{\hbar}\int_x^{x_2} p(x')\,dx' + \frac{\pi}{4}\right)$$

$\pi/4$ "phase shift" from each soft wall.

Thus for one "soft wall" $\int_{x_1}^{x_2} p(x)\,dx = (n-\frac{1}{4})\pi\hbar$, x_1, x_2 turning points $(E=V(x))$

for two "soft walls" $= (n-\frac{1}{2})\pi\hbar$

Half-harmonic oscillator

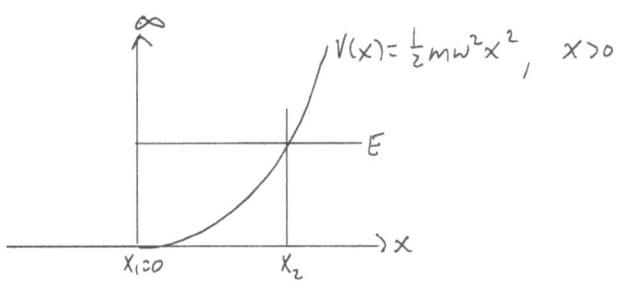

The (unknown) energy E determines the RHS turning point:

$$E = V(x) = \tfrac{1}{2} m \omega^2 x_2^2 \longrightarrow x_2 = \tfrac{1}{\omega}\sqrt{\tfrac{2E}{m}}$$

Then $p(x) = \sqrt{2m(E-V(x))} = \sqrt{2m(\tfrac{1}{2}m\omega^2 x_2^2 - \tfrac{1}{2}m\omega^2 x^2)} = m\omega \sqrt{x_2^2 - x^2}$

$$\int_0^{x_2} m\omega \sqrt{x_2^2 - x^2}\, dx \xrightarrow[\tfrac{\pi}{2} > \theta > 0]{x = x_2 \cos\theta} \int_{\pi/2}^{0} m\omega \sqrt{x_2^2 - x_2^2 \cos^2\theta}\,(-x_2 \sin\theta\, d\theta) = m\omega x_2^2 \int_0^{\pi/2} \overset{\tfrac{1}{2}(1-\cos 2\theta)}{\sin^2\theta}\, d\theta$$

$$= m\omega x_2^2 \left(\tfrac{\theta}{2} - \tfrac{1}{4}\sin 2\theta\right)\Big|_0^{\pi/2} = \tfrac{\pi m \omega}{4}\left(\tfrac{2E}{m\omega^2}\right) = \tfrac{\pi E}{2\omega} = (n-\tfrac{1}{4})\hbar\pi \longrightarrow E = \hbar\omega(2n - \tfrac{1}{2}),\ n=1,2,\dots$$

C.f. exact result (odd eigenfunction solutions of full H.O., $E = \hbar\omega(n - \tfrac{1}{2})$, $n \to 2n$)

Coulomb potential

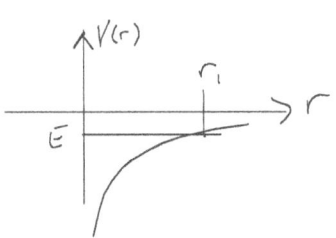

For $\ell = 0$, there is no centrifugal barrier at origin, so we can use origin as turning point, treating it like a hard wall:

$$\int_0^{r_1} \sqrt{2m(E - V(r))}\, dr = (n - \tfrac{1}{4})\pi\hbar, \quad E = V(r_1) = \tfrac{-e^2}{4\pi\varepsilon_0 r_1} \longrightarrow r_1 = \tfrac{-e^2}{4\pi\varepsilon_0 E}$$

$$\int_0^{r_1} \sqrt{2m(E + \tfrac{e^2}{4\pi\varepsilon_0 r})}\, dr = \sqrt{2mE} \int_0^{r_1} \sqrt{1 + (\tfrac{e^2}{4\pi\varepsilon_0 E})\tfrac{1}{r}}\, dr = \sqrt{2mE} \int_0^{r_1} \sqrt{1 - r_1/r}\, dr = \sqrt{2mE}\int_0^{r_1} \tfrac{\sqrt{r^2 - r r_1}}{r}\, dr$$

Now we just have to carry out the integration...

Integral evaluation via trigonometric substitution

Substitute $r = r_1 \sin^2\theta$, where $0 < \theta < \frac{\pi}{2}$. Then, infinitesimal $dr = 2r_1 \sin\theta \cos\theta \, d\theta$.

$$\phi(r_1) = \sqrt{-2mE} \int_0^{\pi/2} \frac{\sqrt{r_1^2 \sin^2\theta - r_1^2 \sin^4\theta}}{r_1 \sin^2\theta} 2r_1 \sin\theta \cos\theta \, d\theta = \sqrt{-2mE} \int_0^{\pi/2} 2r_1 \cos^2\theta \, d\theta,$$

where we write the numerator as $r_1 \sin\theta \sqrt{1 - \sin^2\theta} = r_1 \sin\theta \cos\theta$.

With $\cos^2\theta = \frac{1}{2}(1 + \cos 2\theta)$, we have:

$$= 2r_1 \sqrt{-2mE} \left(\frac{\theta}{2} + \frac{\sin 2\theta}{4}\right)\Big|_0^{\pi/2} = \frac{\pi}{2} \frac{e^2}{4\pi\varepsilon_0 E} \sqrt{-2mE} = -\frac{\pi}{2}\sqrt{-2m}\left(\frac{e^2}{4\pi\varepsilon_0}\right)\frac{1}{\sqrt{E}} = (n - \tfrac{1}{4})\hbar\pi$$

$$E = -\frac{m}{2\hbar^2}\left(\frac{e^2}{4\pi\varepsilon_0}\right)^2 \frac{1}{(n - \tfrac{1}{4})^2} = -\frac{1}{2}mc^2\left(\frac{e^2}{4\pi\varepsilon_0 \hbar c}\right)^2 \frac{1}{(n - \tfrac{1}{4})^2} = -\frac{1}{2}mc^2 \alpha^2/(n - \tfrac{1}{4})^2$$

For large n, $E \approx -\frac{1}{2}\frac{mc^2 \alpha^2}{n^2}$, the exact result! In light of the correspondence principle, we see where the name "semiclassical approximation" comes from!

Two-particle wavefunction

- By adding terms dependent on coordinates $\vec{r}_1, \chi_1$ (for particle #1) and $\vec{r}_2, \chi_2$ (for particle #2) [χ = spin] to the Hamiltonian, we can assemble a Schrödinger equation for the two-particle wave function $\Psi(\vec{r}_1, \chi_1, \vec{r}_2, \chi_2, t)$.

- Now imagine swapping labels $1 \leftrightarrow 2$. Clearly, doing this twice must give us back the original wavefunction. But we could have two possibilities: $\Psi(\vec{r}_2, \chi_2, \vec{r}_1, \chi_1, t) = \pm \Psi(\vec{r}_1, \chi_1, \vec{r}_2, \chi_2, t)$, without any consequence for observables!

- Half-integer spin antisymmetric "fermions" (electrons, protons, neutrons, etc) have $(-)$ and integer-spin symmetric "bosons" (photons, gravitons, Higgs, etc) have $(+)$. This rule (the "spin-statistics Theorem") is a constraint provided by relativity.

- Multi-fermion wavefunctions thus cannot simply be products of single-particle wavefunctions.

- Superpositions of product states can incorporate this exchange symmetry <u>and</u> indistinguishability!

$$\Psi_{ab}(\vec{r}_1, \chi_1, \vec{r}_2, \chi_2) = \frac{1}{\sqrt{2}}\left(\psi_a(\vec{r}_1, \chi_1)\psi_b(\vec{r}_2, \chi_2) \underset{\text{fermions}}{\overset{\text{bosons}}{\pm}} \psi_b(\vec{r}_1, \chi_1)\psi_a(\vec{r}_2, \chi_2)\right)$$

- Note if $a = b$, fermion $\Psi_{ab}(\vec{r}_1, \chi_1, \vec{r}_2, \chi_2) = 0$! ⟶ "Pauli exclusion"

Observables: average interparticle distance

Spatial exchange symmetry has consequences for particle-particle interactions. Consider mean distance between particles: $\sqrt{\langle(x_1-x_2)^2\rangle} = \sqrt{\langle x_1^2\rangle + \langle x_2^2\rangle - 2\langle x_1 x_2\rangle}$ for two indistinguishable particles in different orbital states (a and b)

$$\psi_{ab}(x_1, x_2) = \frac{1}{\sqrt{2}}\left[\psi_a(x_1)\psi_b(x_2) \pm \psi_b(x_1)\psi_a(x_2)\right]$$

The first two expectation values are independent of exchange symmetry, e.g.:

$$\langle x_1^2\rangle = \iint \psi_{ab}^*(x_1, x_2)\, x_1^2\, \psi_{ab}(x_1, x_2)\, dx_1\, dx_2$$

$$= \iint \frac{1}{\sqrt{2}}\left[\psi_a^*(x_1)\psi_b^*(x_2) \pm \psi_b^*(x_1)\psi_a^*(x_2)\right] x_1^2 \frac{1}{\sqrt{2}}\left[\psi_a(x_1)\psi_b(x_2) \pm \psi_b(x_1)\psi_a(x_2)\right] dx_1\, dx_2$$

$$= \frac{1}{2}\Big[\int \psi_a^*(x_1)\, x_1^2\, \psi_a(x_1)\, dx_1 \underbrace{\int \psi_b^*(x_2)\psi_b(x_2)\, dx_2}_{1}$$

$$+ \int \psi_b^*(x_1)\, x_1^2\, \psi_b(x_1)\, dx_1 \underbrace{\int \psi_a^*(x_2)\psi_a(x_2)\, dx_2}_{1}$$

$$\pm \int \psi_a^*(x_1)\, x_1^2\, \psi_b(x_1)\, dx_1 \underbrace{\int \psi_b^*(x_2)\psi_a(x_2)\, dx_2}_{0}$$

$$\pm \int \psi_b^*(x_1)\, x_1^2\, \psi_a(x_1)\, dx_1 \underbrace{\int \psi_a^*(x_2)\psi_b(x_2)\, dx_2}_{0}\Big] = \frac{1}{2}\left[\langle x^2\rangle_a + \langle x^2\rangle_b\right]$$

Calculation of $\langle x_2^2\rangle$ of course yields the same result bc. of indistinguishability.

"Exchange force"

Now for last term in $\sqrt{\langle(x_1-x_2)^2\rangle} = \sqrt{\langle x_1^2\rangle + \langle x_2^2\rangle - 2\langle x_1 x_2\rangle}$:

$$-2\langle x_1 x_2\rangle = -2\iint \frac{1}{\sqrt{2}}\left[\psi_a^*(x_1)\psi_b^*(x_2) \pm \psi_b^*(x_1)\psi_a^*(x_2)\right] x_1 x_2 \frac{1}{\sqrt{2}}\left[\psi_a(x_1)\psi_b(x_2) \pm \psi_b(x_1)\psi_a(x_2)\right] dx_1\, dx_2$$

$$= -\Big[\int \psi_a^*(x_1)\, x_1\, \psi_a(x_1)\, dx_1 \int \psi_b^*(x_2)\, x_2\, \psi_b(x_2)\, dx_2$$

$$+ \int \psi_b^*(x_1)\, x_1\, \psi_b(x_1)\, dx_1 \int \psi_a^*(x_2)\, x_2\, \psi_a(x_2)\, dx_2$$

$$\pm \int \psi_a^*(x_1)\, x_1\, \psi_b(x_1)\, dx_1 \int \psi_b^*(x_2)\, x_2\, \psi_a(x_2)\, dx_2$$

$$\pm \int \psi_b^*(x_1)\, x_1\, \psi_a(x_1)\, dx_1 \int \psi_a^*(x_2)\, x_2\, \psi_b(x_2)\, dx_2\Big]$$

$$= -\left[\langle x\rangle_a \langle x\rangle_b + \langle x\rangle_b \langle x\rangle_a \pm \langle x\rangle_{ab}\langle x\rangle_{ba} \pm \langle x\rangle_{ba}\langle x\rangle_{ab}\right] = -2\langle x\rangle_a \langle x\rangle_b \mp 2|\langle x\rangle_{ab}|^2$$

Thus, $\sqrt{\langle(x_1-x_2)^2\rangle} = \sqrt{\langle x^2\rangle_a + \langle x^2\rangle_b - 2\langle x\rangle_a \langle x\rangle_b \mp 2|\langle x\rangle_{ab}|^2}$

- So exchange-symmetric particles (bosons) feel an effective "exchange force" tending to push particles together, and exchange-antisymmetric (fermions) feel force pushing apart!

Molecular bonding

The spin-statistics theorem demands that the <u>total</u> wavefunction of spin-1/2 electrons is anti-symmetric under exchange. So with spin, even fermions can form the <u>spatially</u> symmetric wavefunction if the spin part is antisymmetric: two-spin singlet $\frac{1}{\sqrt{2}}(|\uparrow\downarrow\rangle - |\downarrow\uparrow\rangle)$

This is what happens in e.g. H_2 molecule: the "bonding" state is spin singlet so that the spatial wavefunction is symmetric under exchange, causing exchange force pushing electrons to the interatomic region, attracting positively-charged ions.

<u>bonding</u>
spatially symmetric
spin singlet

<u>anti-bonding</u>
spatially antisymmetric
spin triplet

Example including Coulomb interaction

Ignoring Pauli exclusion, indistinguishability, and Coulomb repulsion, two electrons in the same 1-D QW potential have total energy $E = \frac{\hbar^2 \pi^2}{2ma^2}(n_1^2 + n_2^2)$, $n_{1,2} = 1, 2, \ldots$

Therefore, including spin up/down for each, we have:

$|n_1, n_2\rangle$	degeneracy
$|1,1\rangle$	4
$|1,2\rangle$	4
$|2,1\rangle$	4

ground: $|1,1\rangle$
1st excited: $|1,2\rangle, |2,1\rangle$

$(|\uparrow\uparrow\rangle, |\downarrow\downarrow\rangle, |\uparrow\downarrow\rangle, |\downarrow\uparrow\rangle)$

— (8) 1st excited
— (4) ground

When we include Pauli exclusion + indistinguishability, many of these degenerate states are eliminated due to overall exchange antisymmetry constraint.

state	degeneracy	
$|1,1\rangle$	1	("singlet") $\frac{1}{\sqrt{2}}(|\uparrow\downarrow\rangle - |\downarrow\uparrow\rangle)$
$\frac{1}{\sqrt{2}}(|1,2\rangle + |2,1\rangle)$	1	("singlet")
$\frac{1}{\sqrt{2}}(|1,2\rangle - |2,1\rangle)$	3	("triplet": $|\uparrow\uparrow\rangle, |\downarrow\downarrow\rangle, \frac{1}{\sqrt{2}}(|\uparrow\downarrow\rangle + |\downarrow\uparrow\rangle)$)

QW + Coulomb
(4) — (1) singlet
     — (3) triplet
— (1) ground

Including Coulomb interaction, triplet becomes lowest excited state because spatial exchange antisymmetry maximizes interelectron distance and minimizes electrostatic energy.

Multi-electron atom: Coulomb interactions

Example: Neutral Helium, two electrons.
Ignoring Coulomb interactions, ground state is just product of $Z=2$ one-electron atom ground state wavefunctions, which must be spatially symmetric (since same n,l,m) with spin singlet.
Energy is just sum: $E_1^{He} = 2^2 E_1^H + 2^2 E_1^H = -8 \cdot 13.6 eV \sim -109 eV$
Coulomb repulsion will add to this energy. Actual $E_1^{He} \sim -79 eV$

- Excited states can form spatially symmetric or antisymmetric superpositions of the single-particle ground state and excited states, with nomenclature "parahelium": singlet (spin 0) / "orthohelium": triplet (spin 1)
The singlet will have attractive exchange force, increasing Coulomb interaction.
Therefore, orthohelium is energetically favored.

Three electrons and more: Slater determinant

Consider a system of three electrons (e.g. neutral lithium): $\{n, l, m_l, m_s\}$
Pauli exclusion: each electron in its own state (a, b, c)

Impose indistinguishability + exchange antisymmetry:

$\psi_a(r_1)\psi_b(r_2)\psi_c(r_3)$, $\psi_b(r_1)\psi_c(r_2)\psi_a(r_3)$ etc.

$\frac{1}{\sqrt{6}}(|abc\rangle + |bca\rangle + |cab\rangle - |cba\rangle - |acb\rangle - |bac\rangle)$

(all permutations of three objects)

Same as a determinant:

$$\frac{1}{\sqrt{6}} \begin{vmatrix} \psi(r_1) & a & b & c \\ \psi(r_2) & a & b & c \\ \psi(r_3) & a & b & c \end{vmatrix} = \frac{1}{\sqrt{6}} \begin{pmatrix} |abc\rangle - |acb\rangle \\ -|bac\rangle + |bca\rangle \\ +|cab\rangle - |cba\rangle \end{pmatrix}$$

For $N > 3$, some states are not just cyclic permutations of abcd...
In general, for N electrons we have the "Slater determinant"

$$\psi(r_1, r_2, \ldots r_N) = \frac{1}{\sqrt{N!}} \begin{vmatrix} \psi_1(r_1) & \psi_2(r_1) & \cdots & \psi_N(r_1) \\ \psi_1(r_2) & \psi_2(r_2) & \cdots & \psi_N(r_2) \\ \vdots & & & \\ \psi_1(r_N) & \psi_2(r_N) & \cdots & \psi_N(r_N) \end{vmatrix}$$

Hund's rules → "rituals" :)

Generalizing from 2-electron atom to higher Z, we have a heuristic explanation for "Hund's rule" #1:

Due to exchange symmetry-induced effects on electron-electron Coulomb repulsion, the lowest energy configuration will maximize total spin $"S" = \sum m_s$

"Hund's rule" #2 tells us that the lowest energy state is the one that maximizes total orbital angular momentum $"L" = \sum m_\ell$. This is often explained using classical arguments of questionable validity, i.e. that electrons orbiting in the same direction (same sign of m_ℓ's) will cross paths less often and minimize Coulomb interaction.

"Hund's rule" #3 says that for atoms with less than half-filled shell (n,ℓ), the lowest energy state minimizes total angular momentum $"J"$. This is heuristically due to oppositely-oriented L and S, so that $\vec{S}\cdot\vec{L} < 0$ and spin-orbit interaction energy is negative.

Hund's rule example

Consider all possible configurations of m_ℓ and m_s allowed by Pauli exclusion principle. Then, apply Hund's rules in order, until only one remains. $(\ell=0)="s"$, $(\ell=1)="p"$, $(\ell=2)="d"$, $(\ell=3)="f"$, ...

Example: Ti ($Z=22$): $1s^2 2s^2 2p^6 3s^2 3p^6 4s^2 3d^2$

How to fill the partially occupied 3d ($\ell=2$) states?

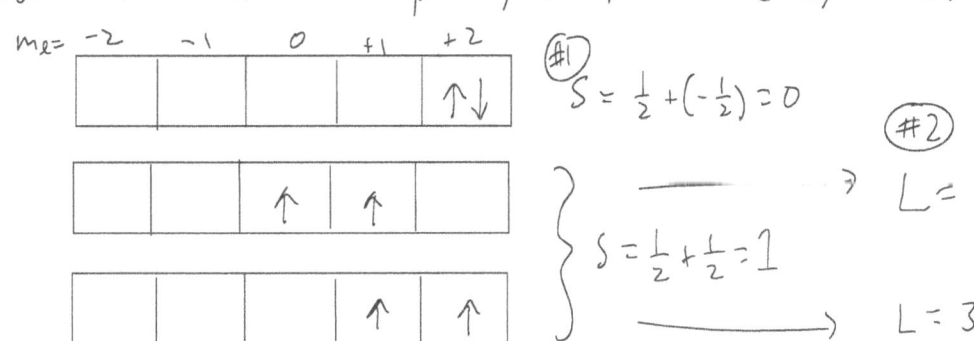

#1 $S = \frac{1}{2} + (-\frac{1}{2}) = 0$

#2 $L=1$, $S = \frac{1}{2} + \frac{1}{2} = 1$ → $L=3$

#3 $J = L-S = 3-1 = 2$ for half-filled. Using notation: L_J^{2S+1} → capitalize alphabetical notation for ℓ → $^3F_2$

Free electron gas

What if we have an even larger number of electrons, like in a metal or plasma? What is the ground state, consistent with the Pauli exclusion principle?

If the electrons are not confined by any potential, and we ignore electron-electron Coulomb interactions, our Hamiltonian is trivial:

$$-\frac{\hbar^2}{2m}\nabla^2 \psi = E\psi \quad \rightarrow \quad \psi \propto e^{i\vec{k}\cdot\vec{r}} \quad (|\vec{k}| = \sqrt{\tfrac{2mE}{\hbar}})$$

Our states are therefore labeled by $k_x, k_y,$ and k_z components of $\vec{k}$. We can fill states from the lowest energy $(\vec{k}=\vec{0})$, until all electrons in the system are accounted for.

Since $E = \frac{\hbar^2}{2m}(k_x^2 + k_y^2 + k_z^2)$, the surface of the volume of filled states in $\vec{k}$-space is a "Fermi sphere", with volume $\Omega = \frac{4}{3}\pi k_F^3$. The radius k_F ("Fermi wavenumber") depends on the number of filled states and the volume each state excludes.

Density of States

Our free electron gas is not confined by any external potential, and so has infinite volume V and infinite total # of electrons N. However, the electron density $n = \frac{N}{V}$ is fixed. Due to full translational symmetry, we can assert arbitrary periodicity.

For example, cut space into identical cubes: $\longrightarrow$

Now, apply periodic boundary conditions: $\psi(\vec{r}) = \psi(\vec{r}+L_x\hat{x}) = \psi(\vec{r}+L_y\hat{y}) = \psi(\vec{r}+L_z\hat{z})$

Using our planewave eigenfunctions, we have eg. $e^{i\vec{k}\cdot\vec{r}} = e^{i\vec{k}\cdot(\vec{r}+L_x\hat{x})} = e^{i\vec{k}\cdot\vec{r}} e^{i\vec{k}\cdot L_x\hat{x}}$

So, $\vec{k}\cdot L_x\hat{x} = k_x L_x = 2\pi \cdot$ integer, and spacing of states along k_x is $\Delta k_x = \frac{2\pi}{L_x}$.

In 3d $\vec{k}$ space, $\Delta^3 k = \frac{(2\pi)^3}{V}$ (where $V \leftarrow L_x L_y L_z$) per state labeled by k_x, k_y, k_z, and half as much when including spin (up/down) degeneracy.

For Fermi sphere w/ volume $\frac{4}{3}\pi k_F^3$, $\quad N = \frac{\frac{4}{3}\pi k_F^3}{\frac{(2\pi)^3}{2V}} \rightarrow n = \frac{k_F^3}{3\pi^2}$

Properties of a "typical" Fermi sphere

In a typical monovalent metal $n \sim 10^{23} \text{ cm}^{-3}$

"Fermi wavenumber" $K_F = (3\pi^2 n)^{1/3} \sim 10^8 \text{ cm}^{-1}$

On the surface of the Fermi sphere,

"Fermi velocity" $v_F = \frac{\hbar K_F}{m} = \frac{6.6 \times 10^{-16} \text{ eV·s} \cdot 10^8 \text{ cm}^{-1}}{5 \times 10^5 \text{ eV} / 9 \times 10^{20} \frac{\text{cm}^2}{\text{s}^2}} \sim 10^8 \frac{\text{cm}}{\text{s}}$

Compare to classical velocity @ RT: $K_B T \sim \frac{1}{2} m v^2 \Rightarrow v \sim \sqrt{\frac{2 K_B T}{m}} = c \sqrt{\frac{1/40 \text{ eV}}{5 \times 10^5 \text{ eV}}} \sim 10^{-4} c \cong 10^6 \frac{\text{cm}}{\text{s}}$

"Fermi energy" $E_F = \frac{\hbar^2 K_F^2}{2m} = \frac{(6.6 \times 10^{-16} \text{ eV·s})^2 (10^8 \text{ cm}^{-1})^2}{2 \cdot 5 \times 10^5 \text{ eV}/c^2} \sim 1-10 \text{ eV}$

"Fermi Temperature" $T_F = \frac{E_F}{K_B} = \frac{10 \text{ eV}}{1 \text{ eV}/40/300\text{K}} \sim 10^5 \text{ K}$ → properties of ground state ($T=0$) are a very good approximation to RT ($T=300$K) behavior!

Total energy

Sum over energy contributions from every electron in Fermi sphere

$$E = 2 \sum_{|\vec{K}| < K_F} \frac{\hbar^2 K^2}{2m}$$

↑ Spin degeneracy

Since $\Delta^3 K = \frac{(2\pi)^3}{V}$, $E = \frac{2 \cdot V}{(2\pi)^3} \sum_{|\vec{K}| < K_F} \frac{\hbar^2 K^2}{2m} \Delta^3 K \longrightarrow \frac{2V}{(2\pi)^3} \int_{\text{Fermi sphere}} \frac{\hbar^2 K^2}{2m} d^3K$

Energy is a function only of $|\vec{K}|$, so we have spherical symmetry:

$d^3 K \to 4\pi K^2 dK$, so $E = \frac{2V}{(2\pi)^3} \int_0^{K_F} \frac{\hbar^2 K^2}{2m} 4\pi K^2 dK = \frac{V \hbar^2 K^5}{2\pi^2 m \cdot 5} \bigg|_0^{K_F} = \frac{\hbar^2 K_F^5}{10 \pi^2 m} V$

Then, Energy density $u = \frac{E}{V} = \frac{\hbar^2 K_F^5}{10 \pi^2 m}$, and average energy per electron

$\frac{u}{n} = \frac{\frac{\hbar^2 K_F^5}{10 \pi^2 m}}{K_F^3 / 3\pi^2} = \frac{3}{10} \frac{\hbar^2 K_F^2}{m} = \frac{3}{5} E_F$

143

Fermi Pressure

$$P = -\frac{\partial E}{\partial V}\Big|_N = -\frac{\partial}{\partial V}\left[\tfrac{3}{5} E_F N\right]\Big|_N = -\tfrac{3}{5} N \frac{\partial}{\partial V}\left[\frac{\hbar^2}{2m}\left(3\pi^2 \frac{N}{V}\right)^{2/3}\right]\Big|_N$$

$$= -\tfrac{3}{5} N \frac{\hbar^2}{2m} \frac{(3\pi^2 \frac{N}{V})^{2/3}}{V}\left(-\tfrac{2}{3}\right) = \tfrac{2}{5} E_F n$$

Assuming $n \sim 10^{23}\,cm^{-3}$, $E_F \sim 10\,eV$, then we have $P \sim 10^{24}\,eV/cm^3$. Convert to more useful units:

$$10^{24}\,eV/cm^3 \cdot \frac{1.6\times 10^{-19}\,J}{eV} \cdot \frac{10^6\,cm^3}{m^3} \simeq 10^{11}\,\frac{N}{m^2} \sim 10^6\,atm. \quad \text{Huge!}$$

What are the physical consequences of such a large degeneracy pressure?

"Compressivity" $= -\frac{dV/dP}{V} \longrightarrow$ "bulk modulus" $B = \frac{1}{\text{compressivity}} = -V\frac{dP}{dV}$

Since $P \propto V^{-5/3}$, $B = -V\left(-\tfrac{5}{3}\frac{P}{V}\right) = \tfrac{5}{3} P \sim \tfrac{2}{3} E_F n \sim 10^6\,atm.$

This confirms the empirically negligible compressibility of metals.

1-d periodic potentials

Now add periodic potential energy from Coulomb interactions w/ ions in crystal lattice:

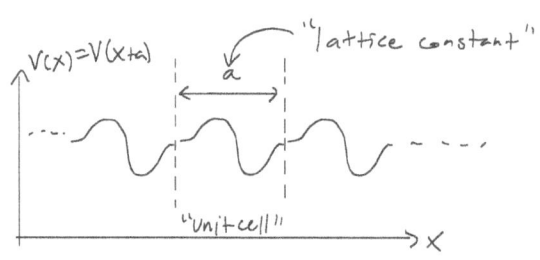

$V(x)$ can be expanded in a Fourier series:

$$V(x) = \sum V_G e^{iGx} \qquad G = G_h = h\frac{2\pi}{a} = 0, \pm\frac{2\pi}{a}, \pm\frac{4\pi}{a}, \ldots \quad (h = \ldots, -1, 0, 1, \ldots)$$

G's are discrete-valued "reciprocal lattice numbers" satisfying periodic symmetry.

We want to solve Schrödinger equation:

$$\left[-\frac{\hbar^2}{2m}\frac{d^2}{dx^2} + V(x)\right]\psi(x) = E\psi(x) \qquad \text{In general, solutions are } \underline{\text{not}} \text{ plane waves!}$$

Bloch theorem

If $V(x)$ is periodic, then Hamiltonian is periodic and commutes with the "displacement operator" D: $Df(x) = f(x+a)$

Therefore, $[D,H] = 0$. We then have simultaneous eigenfunctions $\psi(x)$: $D\psi(x) = \lambda \psi(x)$ so $\psi(x+a) = \lambda \psi(x)$. Since periodicity requires $|\psi(x)|^2 = |\psi(x+a)|^2$, the eigenvalue $\lambda = e^{ika}$. Therefore, $\psi(x+a) = e^{ika}\psi(x)$.

Solutions which satisfy this "Bloch theorem" are "Bloch waves" $\psi(x) = e^{ikx} u(x)$, where $u(x)$ has the same periodicity as the lattice. Obviously, we can expand this envelope function in the same basis as $V(x)$: $u(x) = \sum_G C_G e^{iGx}$.

Schrödinger equation in Fourier space

Substitute Bloch wave $\psi(x) = e^{ikx} u(x)$ into Schrödinger eqn:

$$-\frac{\hbar^2}{2m}\left(e^{ikx} u(x)\right)'' + V(x) e^{ikx} u(x) = E e^{ikx} u(x)$$

$$-\frac{\hbar^2}{2m}\left(u'' + 2iku' - k^2 u\right)e^{ikx} + V(x) e^{ikx} u(x) = E e^{ikx} u(x)$$

$$-\frac{\hbar^2}{2m}\left(\frac{d}{dx} + ik\right)^2 u + V u = E u$$

This is a "Schrödinger equation" for the envelope function $u(x)$.

Substitute $u(x) = \sum_G C_G e^{iGx}$. Then $\frac{d}{dx} \to iG$

$$+\frac{\hbar^2}{2m}\sum_G (k+G)^2 C_G e^{iGx} + \sum_{G'} V_{G'} e^{iG'x} \sum_G C_G e^{iGx} = E \sum_G C_G e^{iGx}$$

Vanishing potential

In the simplest case, $V(x) = 0$ but we keep the imposed discrete periodicity. Then, $V_G = 0$ for all G, and our eigenvalue equation gives

$$+\frac{\hbar^2}{2m} \sum_G (K+G)^2 C_G e^{iGx} = E \sum_G C_G e^{iGx}$$

Since the e^{iGx} functions are orthonormal, the sum is equal term by term:

$$E = +\frac{\hbar^2}{2m}(K+G)^2 \qquad G = 0, \pm \frac{2\pi}{a}, \pm \frac{4\pi}{a} \ldots$$

This result is identical to the case of continuous translation symmetry $\left(E = \frac{\hbar^2 k^2}{2m}\right)$, except now $K \to K+G$!

Free electron bandstructure

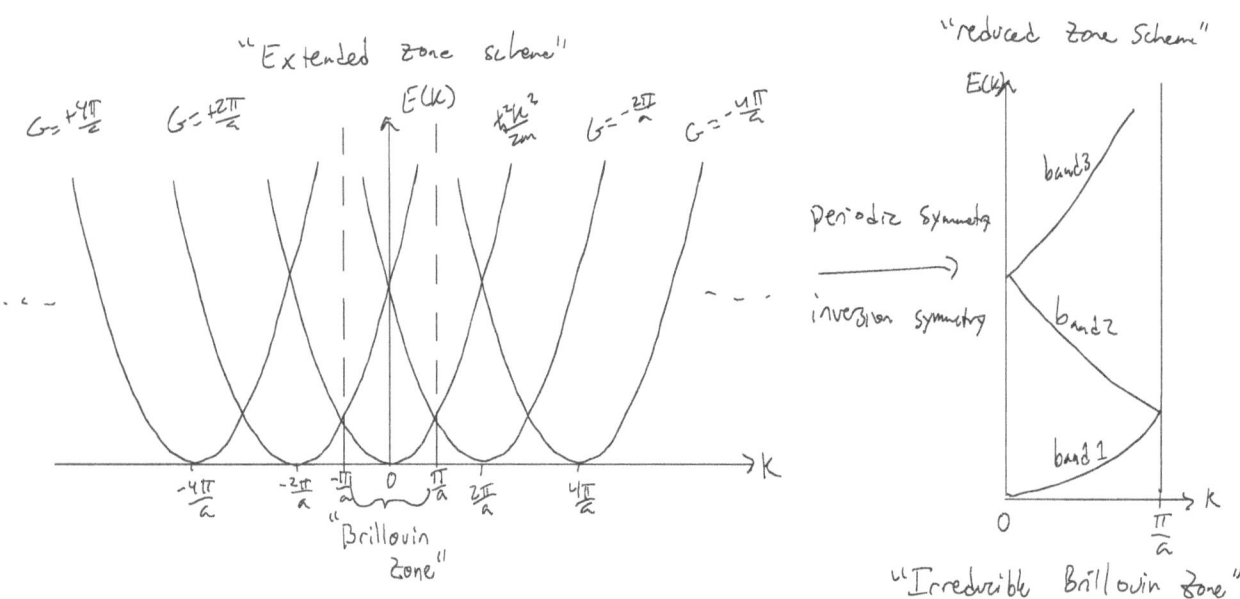

But this <u>still</u> describes a metal! Must understand $V \neq 0$ case!

Nonzero potential

Potential energy term:

$$\sum_{G'} V_{G'} e^{iG'x} \sum_{G} C_G e^{iGx} = \sum_{h}\sum_{h'} V_{h'} C_h e^{i(G_h + G_{h'})x} = \sum_{h}\sum_{h'} V_{h'} C_h e^{iG_{h+h'}x}$$

where $G_h = \frac{2\pi}{a}h$, $G_{h'} = \frac{2\pi}{a}h'$.

Now substitute $h'' \equiv h + h'$:

$$\sum_{h''}\sum_{h'} V_{h'} C_{h''-h'} e^{iG_{h''}x} = \sum_{h}\sum_{h'} V_{h'} C_{h-h'} e^{iG_h x}$$

and our "Schrödinger eqn" becomes

$$\sum_{h}\left[\frac{\hbar^2}{2m}(K+G_h)^2 C_h + \sum_{h'} V_{h'} C_{h-h'}\right] e^{iG_h x} = \sum_{h} E C_h e^{iG_h x}$$

Due to orthonormality of the e^{iGx} basis functions, we have term-by-term equality, so each coeff. must separately satisfy

$$\frac{\hbar^2}{2m}(K+G_h)^2 C_h + \sum_{h'} V_{h'} C_{h-h'} = E C_h \quad \text{(for all } h\text{!)}$$

Infinite matrix eigenvalue problem

$\vdots$

$(h=-1)$ $\ldots V_1 C_{-2} + \frac{\hbar^2}{2m}(K+G_{-1})^2 V_0 C_{-1} + V_{-1} C_0 + V_{-2} C_1 + \ldots = E C_{-1}$

$(h=0)$ $\ldots V_2 C_{-2} + V_1 C_{-1} + \frac{\hbar^2}{2m}(K+G_0)^2 C_0 + V_0 C_0 + V_{-1} C_1 + \ldots = E C_0$

$(h=1)$ $\ldots V_3 C_{-2} + V_2 C_{-1} \quad V_1 C_0 + \frac{\hbar^2}{2m}(K+G_1)^2 C_1 + V_0 C_1 \ldots = E C_1$

$\vdots$

$$\begin{bmatrix} \ddots & & & & & & \\ \cdots V_2 & V_1 & \frac{\hbar^2}{2m}(K+G_0)^2+V_0 & V_{-1} & V_{-2} & \\ & V_2 & V_1 & \frac{\hbar^2}{2m}(K+G_1)^2+V_0 & & \\ & & & & \ddots & \end{bmatrix} \begin{bmatrix} \vdots \\ C_{-1} \\ C_0 \\ C_1 \\ \vdots \end{bmatrix} = E \begin{bmatrix} \vdots \\ C_{-1} \\ C_0 \\ C_1 \\ \vdots \end{bmatrix}$$

147

Finite matrix eigenvalue problem

Cutoff values of G to finite # $\rightarrow$ equivalent to finite spatial resolution. $x \rightarrow \frac{na}{N}$, n is an integer, $1 < n < N$, N is # of Gs.

$e^{i(G_{h'} + G_h)x} = e^{i(h+h')\frac{2\pi}{a}\frac{na}{N}} \rightarrow$ periodic in $h+h'$ w/ period N.

Example: $N = 3$

$$\begin{bmatrix} \frac{\hbar^2}{2m}(k+G_{-1})^2 + V_0 & V_{-1} & V_1 \\ V_1 & \frac{\hbar^2}{2m}(k+G_0)^2 + V_0 & V_{-1} \\ V_{-1} & V_1 & \frac{\hbar^2}{2m}(k+G_1)^2 + V_0 \end{bmatrix} \begin{bmatrix} C_{-1} \\ C_0 \\ C_1 \end{bmatrix} = E \begin{bmatrix} C_{-1} \\ C_0 \\ C_1 \end{bmatrix}$$

Numerical diagonalization

$V = 0$, "free electron" dispersion:

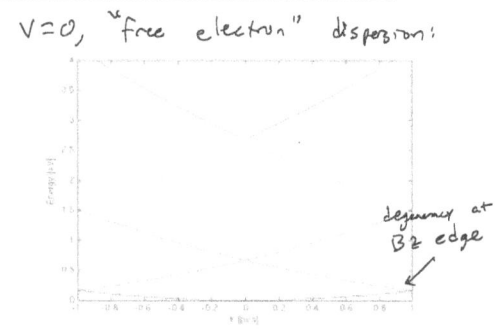

degeneracy at BZ edge

The broken continuous translational symmetry induced by $V(x)$ consequently breaks band degeneracy at BZ edge. This results in "bandgaps", which can be further understood by inspecting the alignment of probability density $\psi^*\psi$ w.r.t. the potential $V(x)$ for k at the BZ edge.

$V = 0.1 \cos \frac{2\pi x}{a}$: (eV) ($a = 1.5$ nm) $\qquad V = 0.5 \cos \frac{2\pi x}{a}$: $\qquad$ wave function density $\psi^*\psi$

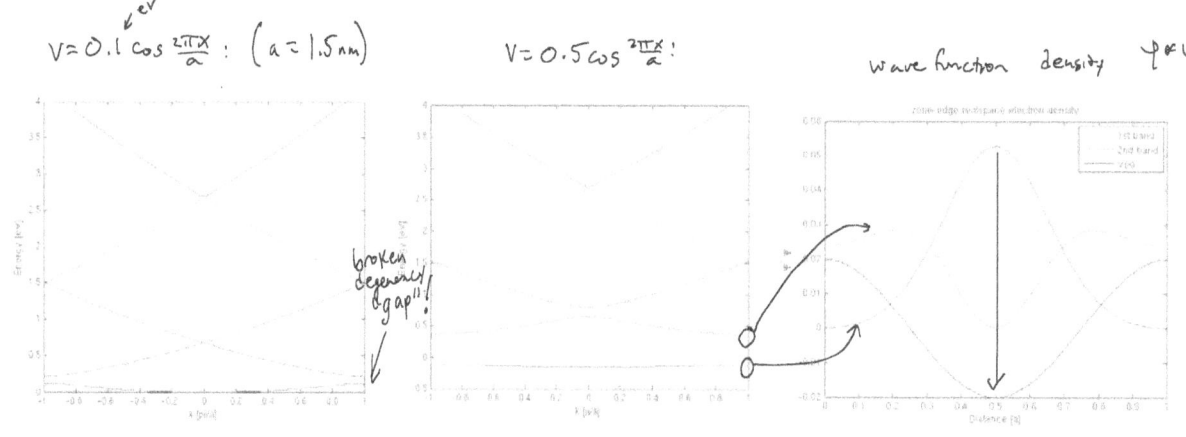

broken degeneracy & gap!

Time-dependent perturbation theory

- If our Hamiltonian is <u>not</u> time-dependent, Schrodinger Eq

$$H\Psi = i\hbar \frac{\partial}{\partial t}\Psi$$

is separable, with stationary states $|\Psi_n(t)\rangle = e^{-i\frac{E_n t}{\hbar}}|\varphi_n\rangle$

- With time-dependence added to Hamiltonian, Schrodinger Eq is not separable and we can no longer express eigenstates as product functions!

- However, we can always expand in an arbitrary set of orthonormal fns multiplied by time-dependent coefficients, assuming our basis remains complete. In other words, only if the time-dependence can be considered a perturbation. Simplest case: two level system, time-independent $H^0|\varphi\rangle = E|\varphi\rangle$ has eigenstates $|\varphi_{a,b}\rangle$ w/ eigenvalues $E_{a,b}$. If time-dependence $H'(t)$ added, use $|\varphi_{a,b}\rangle$ as orthonormal basis to expand $|\Psi\rangle = C_a(t)|\varphi_a\rangle + C_b(t)|\varphi_b\rangle$. $C_{a,b}(t)$ determined by initial conditions and behavior of perturbation $H'(t)$.

Adding perturbation

$$H\Psi = i\hbar \frac{\partial}{\partial t}\Psi, \qquad H = H^0 + H'(t)$$

$$(H^0 + H')(C_a|\varphi_a\rangle + C_b|\varphi_b\rangle) = i\hbar\frac{\partial}{\partial t}(C_a|\varphi_a\rangle + C_b|\varphi_b\rangle)$$

$$C_a H^0|\varphi_a\rangle + C_b H^0|\varphi_b\rangle + C_a H'|\varphi_a\rangle + C_b H'|\varphi_b\rangle = i\hbar(\dot{C}_a|\varphi_a\rangle + \dot{C}_b|\varphi_b\rangle)$$

multiply on left by $\langle\varphi_a|$

$$C_a\langle\varphi_a|H^0|\varphi_a\rangle + C_b\underbrace{\langle\varphi_a|H^0|\varphi_b\rangle}_{0} + C_a\langle\varphi_a|H'|\varphi_a\rangle + C_b\langle\varphi_a|H'|\varphi_b\rangle = i\hbar(\dot{C}_a\underbrace{\langle\varphi_a|\varphi_a\rangle}_{1} + \dot{C}_b\underbrace{\langle\varphi_a|\varphi_b\rangle}_{0})$$

$$E_a C_a + C_a H'_{aa} + C_b H'_{ab} = i\hbar \dot{C}_a$$

Likewise, upon projection onto $\langle\varphi_b|$,

$$E_b C_b + C_b H'_{bb} + C_a H'_{ba} = i\hbar \dot{C}_b$$

Initial conditions

Suppose $C_a(t=0) = 1$, $C_b(t=0) = 0$, and Assume $H'_{aa} = H'_{bb} = 0$

$$\dot{C}_a = -\frac{i}{\hbar}\left(E_a C_a + (0) H'_{ab}\right) \longrightarrow C_a = e^{-\frac{iE_a t}{\hbar}}$$

$$\dot{C}_b = -\frac{i}{\hbar}\left(E_b C_b + (e^{-iE_a t/\hbar}) H'_{ba}\right) \longrightarrow C_b = e^{-iE_b t/\hbar}\left(\frac{-i}{\hbar}\int_0^t H'_{ba}(t') e^{i\frac{E_b - E_a}{\hbar} t'} dt'\right)$$

This is correct to first-order only. We could continue to bootstrap to higher orders by taking C_b and using it in our eqn for $\dot{C}_a$, etc. However, first order is usually sufficient under the usual assumption of weak perturbation $H'(t)$.

Example: spin 1/2 in magnetic field

Static field $\vec{B} = B_z \hat{z}$ so that $H^0 = \mu_B B_z \hat{z} \cdot \vec{\sigma} = \mu_B B_z \sigma_z$

$|\psi_a\rangle = |\downarrow\rangle$, $|\psi_b\rangle = |\uparrow\rangle$, $E_a = -\mu_B B_z$, $E_b = +\mu_B B_z$

Now add time varying field $\vec{B}' = B_x(t)\hat{x}$, so that $H'(t) = \mu_B B_x(t)\hat{x}\cdot\vec{\sigma} = \mu_B B_x(t) \sigma_x$

Note that since $\sigma_x = \begin{bmatrix} 0 & 1 \\ 1 & 0 \end{bmatrix}$, $H'_{aa} = H'_{bb} = 0$ as assumed previously.

$$H = \mu_B \begin{bmatrix} B_z & B_x(t) \\ B_x(t) & -B_z \end{bmatrix}.$$ If $B_x(t) = B_x \cos\omega t$ and the spin is initially in state $|\downarrow\rangle$, then $C_\downarrow(0) = 1$, $C_\uparrow(0) = 0$ and (to first order)

$$C_\uparrow(t) = e^{-i\mu_B B_z t/\hbar}\left(\frac{-i}{\hbar}\int_0^t \mu_B B_x \cos\omega t' \, e^{i\frac{2\mu_B B_z t'}{\hbar}} dt'\right)$$

This is, in general, nonzero, meaning that our perturbation will drive transitions between levels.

Magnetic resonance

$$C_\uparrow(t) = e^{-i\mu_B B_z t/\hbar}\left(-\frac{i}{\hbar}\int_0^t \mu_B B_x \cos\omega t' \, e^{i 2\mu_B B_z t'/\hbar} \, dt'\right)$$

writing $\omega_0 = \frac{2\mu_B B_z}{\hbar}$ and expanding $\cos\omega t' = \frac{e^{i\omega t'} + e^{-i\omega t'}}{2}$

$$= e^{-i\mu_B B_z t/\hbar}\left(-\frac{i\mu_B B_x}{2\hbar}\int_0^t \left(e^{i(\omega_0+\omega)t'} + e^{i(\omega_0-\omega)t'}\right)dt'\right)$$

$$= e^{-i\mu_B B_z t/\hbar}\left(-\frac{\mu_B B_x}{2\hbar}\left[\frac{e^{i(\omega_0+\omega)t}-1}{\omega_0+\omega} + \frac{e^{i(\omega_0-\omega)t}-1}{\omega_0-\omega}\right]\right)$$

For $\omega \simeq \omega_0$, first term negligible.

$$\simeq e^{-i\mu_B B_z t/\hbar}\left(-\frac{\mu_B B_x}{2\hbar}\frac{e^{i(\omega_0-\omega)t/2}}{\omega_0-\omega}\left(e^{i(\omega_0-\omega)t/2} - e^{-i(\omega_0-\omega)t/2}\right)\right)$$

$$= e^{-i\mu_B B_z t/\hbar}\left(-i\frac{\mu_B B_x}{\hbar}\frac{\sin\left((\omega-\omega_0)t/2\right)}{\omega-\omega_0} e^{i(\omega-\omega_0)t/2}\right)$$

Transition Probability

$$P_{\downarrow\to\uparrow}(t) = |C_\uparrow(t)|^2 \simeq \left(\frac{\mu_B B_x}{\hbar}\right)^2 \frac{\sin^2\left((\omega-\omega_0)t/2\right)}{(\omega-\omega_0)^2}$$

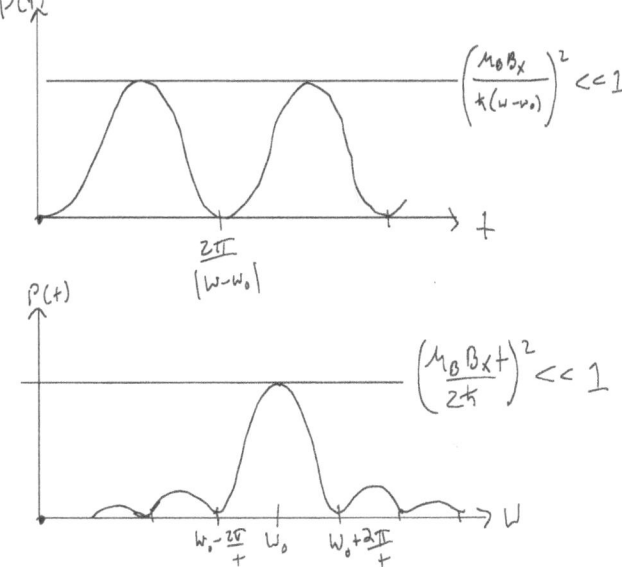

So driving the system at freq. ω causes the transition probability to oscillate at a frequency determined by "detuning" $\Delta\omega = |\omega - \omega_0|$. With stronger driving amplitude B_x, probability grows quadratically. Remember though that this result is valid only for small $P(t)$, where $\frac{\mu_B B_x}{\hbar \Delta\omega} \ll 1$.

Rabi's rotating wave

We can compare the 1st-order perturbation theory approx. to the **exact** result by considering a "rotating" wave, i.e. a circularly-polarized oscillatory magnetic field, $\vec{B}(t) = B'[\cos\omega t\,\hat{x} + \sin\omega t\,\hat{y}]$. Then,

$$H' = \mu_B \vec{B}\cdot\vec{\sigma} = \mu_B B'[\cos\omega t\,\sigma_x + \sin\omega t\,\sigma_y] = \mu_B B'\begin{bmatrix} 0 & \cos\omega t - i\sin\omega t \\ \cos\omega t + i\sin\omega t & 0 \end{bmatrix} = \mu_B B'\begin{bmatrix} 0 & e^{-i\omega t} \\ e^{i\omega t} & 0 \end{bmatrix}$$

Previously, $C_{a,b}$ contained all time dependence, including dynamical phase $e^{-iE_n t/\hbar}$. Making Th.3 standard time-dependence explicit here, then (without approximation):

$$\dot{C}_a = -\frac{i}{\hbar} H'_{ab} e^{-i\omega_0 t} C_b \quad\text{and}\quad \dot{C}_b = -\frac{i}{\hbar} H'_{ba} e^{i\omega_0 t} C_a, \quad\text{where } \omega_0 \equiv \frac{E_b - E_a}{\hbar}$$

For $H'_{ab} = \mu_B B' e^{i\omega t}$, $H'_{ba} = \mu_B B' e^{-i\omega t}$ as above, where $V \equiv 2\mu_B B'$,

$$\dot{C}_a = -\frac{i}{\hbar}\frac{V}{2} e^{i(\omega-\omega_0)t} C_b \quad\text{and}\quad \dot{C}_b = -\frac{i}{\hbar}\frac{V}{2} e^{-i(\omega-\omega_0)t} C_a$$

Decoupling ODEs

Beginning with $\dot{C}_b = \underbrace{-\frac{i}{\hbar}\frac{V}{2} e^{-i(\omega-\omega_0)t}}_{K} C_a$, take time-derivative:

$$\ddot{C}_b = -i(\omega-\omega_0)\dot{C}_b + \left(\frac{\dot{C}_b}{C_a}\right)\dot{C}_a$$

Since $\dot{C}_a = -\frac{i}{\hbar}\frac{V}{2} e^{i(\omega-\omega_0)t} C_b$, $\dot{C}_b \dot{C}_a = -\left(\frac{V}{2\hbar}\right)^2 C_a C_b$ and

$$\ddot{C}_b = -i(\omega-\omega_0)\dot{C}_b - \frac{V^2}{4\hbar^2} C_b$$

- Note similarity to LCR circuit, mass on spring w/ friction, etc: damped resonance!
- This is a 2nd-order ODE w/ constant coefs, with solution $\propto e^{\lambda t}$. Its characteristic quadratic equation $\lambda^2 + i(\omega-\omega_0)\lambda + \frac{V^2}{4\hbar^2} = 0$ is solved by the familiar quadratic formula

$$\lambda = \frac{-i(\omega-\omega_0) \pm \sqrt{-(\omega-\omega_0)^2 - \frac{V^2}{\hbar^2}}}{2} = i\left(-\frac{\omega-\omega_0}{2} \pm \omega_r\right), \quad \omega_r \equiv \frac{1}{2}\sqrt{(\omega-\omega_0)^2 + \frac{V^2}{\hbar^2}}$$

Imposing initial conditions

Separating the $e^{\pm i\omega_r t}$ parts into even + odd components, our soln can be written $C_b(t) = e^{-i\frac{\omega-\omega_0}{2}t}(A\cos\omega_r t + B\sin\omega_r t)$, where coefs A + B are determined by imposing initial conditions:

$C_b(t=0) = 0$ so $A=0$. Since $\dot{C}_b = \frac{-iV}{\hbar 2} e^{-i(\omega-\omega_0)t} C_a$,

$C_a(t) = \frac{i 2\hbar}{V} e^{i(\omega-\omega_0)t} \dot{C}_b = \frac{i 2\hbar}{V} e^{i(\omega-\omega_0)t} \left[B\left(i\frac{\omega_0-\omega}{2} e^{i(\omega_0-\omega)t/2}\sin\omega_r t + \omega_r \cos\omega_r t \, e^{i(\omega_0-\omega)\frac{t}{2}}\right)\right]$

$= \frac{i 2\hbar}{V} e^{i(\omega-\omega_0)t/2} B\left(i\frac{\omega_0-\omega}{2}\sin\omega_r t + \omega_r \cos\omega_r t\right)$

Imposing $C_a(t=0) = 1$, we see $\frac{i 2\hbar}{V} B \omega_r = 1 \Rightarrow B = \frac{-iV}{2\hbar\omega_r}$

Now we have our unique, exact solution, without approximation!

Limiting behavior

Assembling our solution, we have
$$C_a(t) = e^{i(\omega-\omega_0)t/2}\left(\cos\omega_r t + i\frac{\omega_0-\omega}{2\omega_r}\sin\omega_r t\right)$$
$$C_b(t) = \frac{-i}{2\hbar\omega_r} V e^{i(\omega_0-\omega)t/2}\sin\omega_r t$$

For weak driving conditions $V \ll \hbar|(\omega-\omega_0)|$, $\omega_r = \frac{1}{2}\sqrt{(\omega-\omega_0)^2 + \frac{V^2}{\hbar^2}} \approx \frac{\omega-\omega_0}{2}$

Then, $C_b(t) \approx \frac{-iV}{2\hbar \frac{\omega-\omega_0}{2}} e^{i(\omega_0-\omega)t/2}\sin\frac{\omega-\omega_0}{2}t \longrightarrow |C_b(t)|^2 \approx \frac{V^2}{\hbar^2}\frac{\sin^2\frac{\omega-\omega_0}{2}t}{(\omega-\omega_0)^2}$

This is the same result from 1st-order perturbation theory! Now, however, we can examine our prediction in the regime where perturbation theory fails, $\hbar|(\omega-\omega_0)| \ll V$, occurring at resonance $\omega = \omega_0$ where $\omega_r = \frac{V}{2\hbar}$ so $|C_b(t)|^2 = \sin^2\omega_r t$, $|C_a(t)|^2 = \cos^2\omega_r t$. Note these are bounded by unity, unlike our 1st-order approximation!

Energy flow at resonance

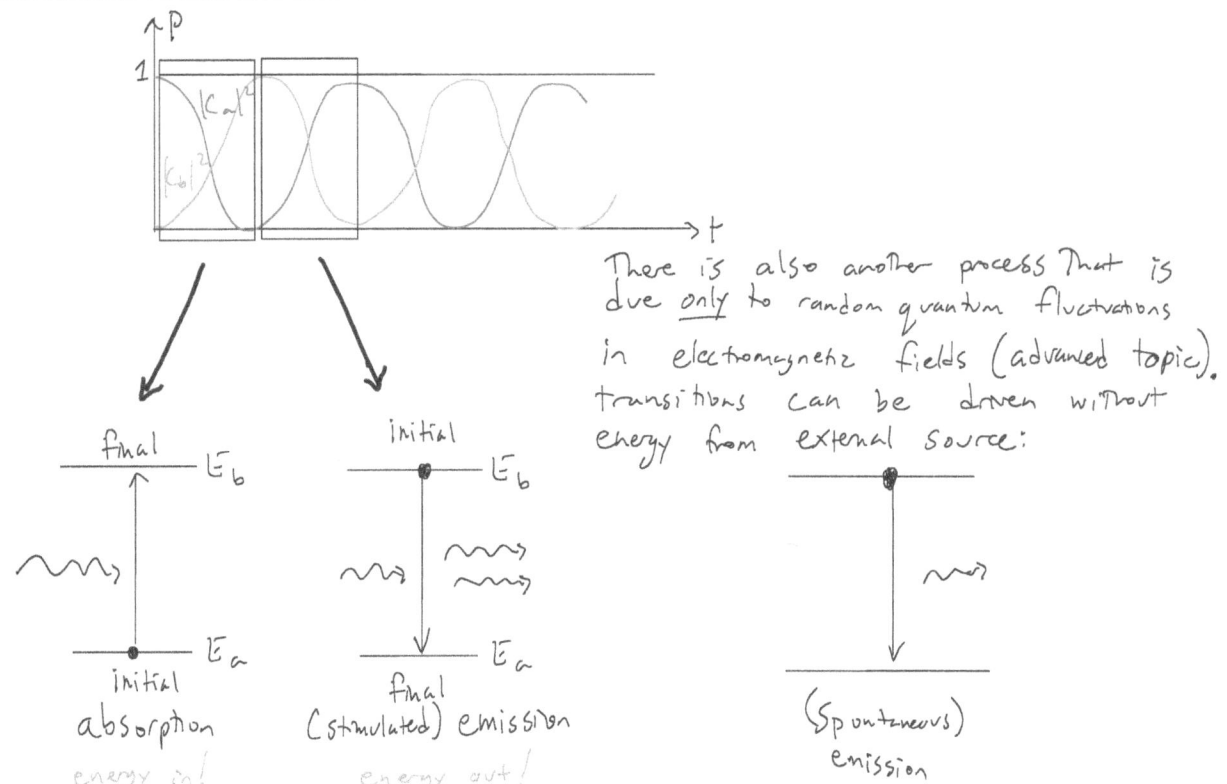

There is also another process that is due <u>only</u> to random quantum fluctuations in electromagnetic fields (advanced topic). Transitions can be driven without energy from external source:

Electromagnetic perturbation

- We want to include this energy exchange w/ perturbing fields in description of our 2-level system. For this purpose, consider orbital states of electrons in atoms, rather than spin-$\frac{1}{2}$.

- We previously saw that (subject to selection rules) transitions from excited states result in radiative emission (photons). The electric field will, in turn, perturb the electron state via absorption.

- Oscillating electric field along $\hat{z}$ corresponds to perturbing time-dependent Hamiltonian $H'(t) = eE_0 \hat{z} \cos\omega t$

- In spherically-symmetric Coulomb potential, all electron orbital states have definite inversion symmetry, so $\langle \varphi_a | H' | \varphi_a \rangle = 0$ (diagonal elements). However, $\langle \varphi_a | H' | \varphi_b \rangle \neq 0$ for different states (off-diagonal elements)

- From 1st order pert. thry, $P_{a \to b}(t) = P_{b \to a}(t) = \left(\frac{eE_0 |\langle \varphi_b | z | \varphi_a \rangle|}{\hbar} \right)^2 \frac{\sin^2\left(\frac{\omega_0 - \omega}{2}\right)t}{(\omega_0 - \omega)^2}$

Fermi's Golden Rule

- What about incoherent radiation over a distribution of frequencies?

- "Fermi's Golden Rule": $\Gamma_{a\to b} = \frac{2\pi}{\hbar} |\langle a|H'|b\rangle|^2 \overbrace{D(E_b)}^{\text{density of states}}$ (follows from $\lim_{t\to\infty} \frac{\sin^2 \omega t}{\omega^2 t^2} = \pi \delta(\omega)$)

To lowest order, consider only 1 state at resonance.

Then $|\langle a|H'|b\rangle|^2 = e^2 E_0^2 |\langle z\rangle_{ab}|^2 \langle \cos^2 \omega t\rangle^2 = e^2 |\langle z\rangle_{ab}|^2 \left(\frac{\epsilon_0 E_0^2}{2}\right) \cdot \frac{2}{\epsilon_0} \cdot \left(\frac{1}{2}\right)^2$

So $\Gamma_{a\to b} = \frac{2 \cdot 2\pi}{4\hbar} e^2 |\langle z\rangle_{ab}|^2 \, u \cdot \frac{1}{\hbar \omega_0} = \frac{\pi}{\epsilon_0 \hbar^2} e^2 |\langle z\rangle_{ab}|^2 \rho(\omega_0)$

where $\frac{u}{\omega_0} \equiv \rho(\omega_0)$ energy density / spectral density (units $\frac{eV \cdot s}{cm^3}$) is essentially assumption of unity relative bandwidth.

- Note that if energy density is distributed among 3 directions, transition probability is $1/3$ this amount.

An aside on Boltzmann factor

To describe the photon-atom system in equilibrium, we need to know occupation probability at finite temperature T. Consider an interaction between two atoms where energy is exchanged, causing electronic transitions:

Energy conservation: $E_1 + E_2 = E_3 + E_4$

Occupation probability conservation / reversibility: $f(E_1,T) f(E_2,T) = f(E_3,T) f(E_4,T)$

What is occupation probability function $f(E,T)$, where T is temperature, subject to intuitive constraints?:

$$\frac{d}{dT} \frac{f(E+\Delta E, T)}{f(E)} > 0, \quad \text{and } T \text{ is proportional to thermal energy}$$

$f(E,T) = A e^{-E/k_B T}$ where k_B has units of energy/temperature. "Maxwell-Boltzmann"

Fermi-Dirac

To include Pauli exclusion, we have to make sure the final states are <u>unoccupied</u>:

$$f(E_1)f(E_2)\left[1-f(E_3)\right]\left[1-f(E_4)\right] = f(E_3)f(E_4)\left[1-f(E_1)\right]\left[1-f(E_2)\right]$$

In this case, $f(E,T) = \dfrac{1}{1+Ae^{E/kT}} = \dfrac{1}{1+e^{(E-E_F)/kT}}$ satisfies the constraints. This is known as the "Fermi-Dirac" distribution.

Atom-photon ensemble at equilibrium

E_b ——— N_b Rate equation: "detailed balance"
$\rho(\omega) \leftrightarrow \hbar\omega_0$ $\dfrac{dN_b}{dt} = N_a B \rho(\omega_0) - N_b B \rho(\omega_0) - N_b A$
E_a ——— N_a (absorption)   (stimulated   (spontaneous
                                      emission)   emission)

In steady-state, $\dfrac{d}{dt}N_b = 0$ and $\rho(\omega_0) = \dfrac{A}{\frac{N_a}{N_b}B - B}$

Assume that photons are in thermal equilibrium w/ atoms. Then $\dfrac{N_a}{N_b} = e^{\hbar\omega_0/kT}$

So $\rho(\omega_0) = \dfrac{A}{B\left(e^{\hbar\omega_0/kT}-1\right)}$: compare to "blackbody" radiation $\rho(\omega) = \dfrac{\hbar}{\pi^2 c^3}\dfrac{\omega^3}{e^{\hbar\omega/kT}-1}$

Therefore,

$\dfrac{A}{B} = \dfrac{\hbar\omega_0^3}{\pi^2 c^3}$ (from perturbation thry) $B = \dfrac{e^2 \langle z \rangle_{ab}^2 \pi}{3\epsilon_0 \hbar^2}$ $\Rightarrow$ $A = \dfrac{\omega_0^3 e^2 \langle z \rangle_{ab}^2}{3\pi \epsilon_0 \hbar c^3} = \dfrac{4}{3}\dfrac{\omega_0^3}{c^2}\left(\dfrac{e^2}{4\pi\epsilon_0 \hbar c}\right)\langle z \rangle_{ab}^2 = \dfrac{4}{3}\dfrac{\omega_0^3 \alpha}{c^2}\langle z \rangle_{ab}^2$

Estimating A

- In atomic systems, $\langle z \rangle \sim a_0 = \frac{\hbar}{mc\alpha}$, $\omega_0 \sim \frac{E_0}{\hbar} = \frac{1}{2}\frac{mc^2\alpha^2}{\hbar}$

So $A \sim \frac{\omega_0}{c^2}\left(\frac{mc^2\alpha^2}{\hbar}\right)^2 \alpha \left(\frac{\hbar}{mc\alpha}\right)^2 = \omega_0 \alpha^3 \sim 10^{-6} \omega_0$.

Therefore, the spontaneous emission rate is much slower than internal dynamics, validating our application of perturbation theory!

- We can get the same result w/o thermodynamics by equating quantum + classical power transfer:

(quantum radiative emission) $\hbar\omega_0 A = \frac{|\ddot{d}|^2}{6_0 c^3} = \frac{a_0^2 e^2 \omega_0^4}{6_0 c^3}$ (classical dipole radiation)

$$A = \left[\left(\frac{e^2}{\hbar 6_0 c}\right) \frac{a_0^2 \omega_0^2}{c^2}\right]\omega_0 = \alpha \left(\frac{a_0 \omega_0}{c}\right)^2 \omega_0$$

Now, since $\left(\frac{a_0 \omega_0}{c}\right)^2 = \left(\frac{\hbar}{mc\alpha}\frac{mc^2\alpha^2}{\hbar c}\right)^2 = \alpha^2$, $= \alpha^3 \omega_0$ Same as above!

Magnetic transitions

This approach is applicable to any radiative process, in the semiclassical limit.

For example, we can return to spin-$\frac{1}{2}$, which can only radiate by magnetic dipole oscillation. Then, equating spontaneous emission power transfer gives $\hbar\omega_0 A = \frac{\mu_0 \mu_B^2 \omega_0^4}{c^3}$.

$$A = \left[\mu_0 \frac{\mu_B^2 \omega_0^3}{\hbar c^3}\right]\omega_0 = \frac{\left(\frac{e\hbar}{2m}\right)^2 \omega_0^2}{\epsilon_0 \hbar c^5}\omega_0 \simeq \frac{e^2}{\epsilon_0 \hbar c}\frac{\hbar^2 \omega_0^2}{m^2 c^4}\omega_0 = \alpha \cdot \left(\frac{\hbar\omega_0}{mc^2}\right)^2 \omega_0$$

Since $\hbar\omega_0 = g\mu_B B$, spontaneous emission rates for spin up → spin down transitions are suppressed. (Even for $B \sim 10 \mathrm{T}$, $g\mu_B B < 1 \mathrm{meV}$, so $\frac{\hbar\omega_0}{mc^2} \sim 10^{-8}$, at best)

Adiabatic theorem

What happens to an eigenstate if time-dependence is so slow that transitions to other states (which we can calculate w/ time-dep. perturbation theory) can be ignored? Examine Schrödinger equation

$$i\hbar \frac{\partial}{\partial t}\Psi(t) = H(t)\Psi(t)$$

Express solution as an expansion in basis of instantaneous eigenstates:

$$\Psi(t) = \sum_n c_n(t)\psi_n(t) e^{i\theta_n(t)}, \quad \theta_n(t) = -\frac{1}{\hbar}\int_0^t E_n(t')dt'$$

Upon substitution, $i\hbar \sum_n [\dot{c}_n \psi_n + c_n \dot{\psi}_n + i\dot{\theta}_n c_n \psi_n] e^{i\theta_n} = \sum_n c_n (H\psi_n) e^{i\theta_n}$

But, $\dot{\theta}_n = -\frac{E_n}{\hbar}$ and $H\psi_n = E_n \psi_n$, so

$$i\hbar \sum_n [\dot{c}_n \psi_n + c_n \dot{\psi}_n] e^{i\theta_n} + \cancel{\sum_n E_n c_n \psi_n e^{i\theta_n}} = \cancel{\sum_n E_n c_n \psi_n e^{i\theta_n}}$$

$$\sum_n \dot{c}_n \psi_n e^{i\theta_n} = -\sum_n c_n \dot{\psi}_n e^{i\theta_n} \xrightarrow{\text{Project onto } \langle \psi_m |} \dot{c}_m e^{i\theta_m} = -\sum_n c_n \langle \psi_m | \dot{\psi}_n \rangle e^{i\theta_n}$$

$$\dot{c}_m = -c_m \langle \psi_m | \dot{\psi}_m \rangle - \sum_{n \neq m} c_n \langle \psi_m | \dot{\psi}_n \rangle e^{i(\theta_n - \theta_m)}$$

Inner product

It may not be obvious that we can ignore the sum. Examine $\langle \psi_m | \dot{\psi}_n \rangle$, $n \neq m$ by manipulating the instantaneous time-independent Schrödinger equation:

$$\frac{d}{dt}[H\psi_n = E_n \psi_n] \Rightarrow \dot{H}\psi_n + H\dot{\psi}_n = \dot{E}_n \psi_n + E_n \dot{\psi}_n$$

Project onto $\langle \psi_m |$:

$$\langle \psi_m | \dot{H} | \psi_n \rangle + \langle \psi_m | H | \dot{\psi}_n \rangle = \dot{E}_n \cancelto{0}{\langle \psi_m | \psi_n \rangle} + E_n \langle \psi_m | \dot{\psi}_n \rangle$$

So $\langle \psi_m | \dot{H} | \psi_n \rangle = (E_n - E_m)\langle \psi_m | \dot{\psi}_n \rangle \Rightarrow \langle \psi_m | \dot{\psi}_n \rangle = \frac{\langle \dot{H} \rangle_{mn}}{E_n - E_m}$ (E_n non-degenerate)

Our result from above therefore gives

$$\dot{c}_m = -c_m \langle \psi_m | \dot{\psi}_m \rangle - \sum_{n \neq m} c_n \frac{\langle \dot{H} \rangle_{mn}}{E_n - E_m} e^{i(\theta_n - \theta_m)}$$

The denominator further suppresses all terms proportional to our (already small) $\dot{H}$. We can therefore confidently neglect the sum and approximate

$$\dot{c}_m \approx -c_m \langle \psi_m | \dot{\psi}_m \rangle$$

Berry's phase

Solution to this first-order differential equation is given by
$$C_m(t) = C_m(0) e^{i\gamma_m(t)}, \quad \gamma_m(t) = i\int_0^t \langle \varphi_m(t') | \frac{\partial}{\partial t'} \varphi_m(t') \rangle dt'$$

γ is real since
$$\frac{d}{dt}\langle \varphi_m | \varphi_m \rangle = \langle \varphi_m | \dot{\varphi}_m \rangle + \langle \dot{\varphi}_m | \varphi_m \rangle = 2\text{Re}\langle \varphi_m | \dot{\varphi}_m \rangle = 0$$

So if $\langle \varphi_m | \dot{\varphi}_m \rangle$ is imaginary, $i\langle \varphi_m | \dot{\varphi}_m \rangle$ is real, and $e^{i\gamma}$ just gives an additional "geometric" or "Berry's" phase to our complete wavefunction:

$$\Psi_n(t) \cong \varphi_n(t) e^{i\theta_n(t)} e^{i\gamma_n(t)}$$

Example: spin-1/2 in rotating B-field

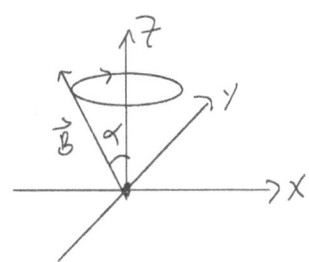

Like Rabi's problem, except now we don't consider $|\uparrow\rangle / |\downarrow\rangle$ along $\hat{z}$, but rather phase of $|+\rangle / |-\rangle$ along $\vec{B}$ at fixed angle $\theta = \alpha$ w/ $\hat{z}$, and azimuthal rotation $\phi = \omega t$.

Our rotating basis are eigenvectors of Zeeman Hamiltonian:
$$H = \frac{g\mu_B}{2} \vec{B} \cdot \vec{\sigma} = \mu_B |B| \left[\cos\alpha\, \sigma_z + \sin\alpha (\cos\omega t\, \sigma_x + \sin\omega t\, \sigma_y) \right] = \mu_B |B| \begin{bmatrix} \cos\alpha & e^{-i\omega t}\sin\alpha \\ e^{i\omega t}\sin\alpha & -\cos\alpha \end{bmatrix}$$

eigenvectors are:
$$|+\rangle = \begin{bmatrix} \cos\alpha/2 \\ e^{i\omega t}\sin\alpha/2 \end{bmatrix}, \quad |-\rangle = \begin{bmatrix} e^{-i\omega t}\sin\alpha/2 \\ -\cos\alpha/2 \end{bmatrix}$$

Geometric phase

$$\gamma(t) = i\int_0^t \langle \varphi(t') | \dot{\varphi}(t') \rangle \, dt'$$

For spin-down:

$$\langle -|\tfrac{d}{dt}|-\rangle = \begin{bmatrix} e^{+i\omega t}\sin\alpha/2, & -\cos\alpha/2 \end{bmatrix} \begin{bmatrix} (-i\omega)e^{-i\omega t}\sin\alpha/2 \\ 0 \end{bmatrix} = -i\omega\sin^2(\alpha/2)$$

So $\gamma_-(t) = i\int_0^t \langle -|\tfrac{d}{dt}|-\rangle \, dt' = \omega t \sin^2(\alpha/2) = \tfrac{\omega t}{2}(1-\cos\alpha)$. Likewise, $\gamma_+(t) = \tfrac{\omega t}{2}(\cos\alpha - 1)$.

In every rotation, $T = \tfrac{2\pi}{\omega}$, Berry's geometric phase is $\gamma_\pm(T) = \pm\pi(\cos\alpha - 1)$

This quantum result is reminiscent of **classical** geometric phase like that observed in precession of the Foucault pendulum due to Earth's rotation:

Daily precession angle determined by solid angle circumscribed:

$$\Omega = 2\pi(1 - \cos\theta_0) \quad \text{c.f. above!}$$

Foucault pendulum

Parametric definition of Berry's phase

We previously defined Berry's phase as $\gamma_n(t) = i\int_0^t \langle \varphi_n(t') | \tfrac{\partial}{\partial t'} \varphi_n(t') \rangle \, dt'$, where we explicitly exploit the time-dependence of the wavefunction. However, that time dependence is caused by parameters of the Hamiltonian changing with time: $H(\vec{R}(t))$, where $\vec{R}(t)$ is a "vector" in a "phase space" defined by the parameter values. Thus, making use of $\tfrac{\partial \varphi}{\partial t} = \tfrac{\partial \varphi}{\partial R}\tfrac{\partial R}{\partial t}$, we can instead write $\gamma_n = i\int_{R_1}^{R_2} \langle \varphi_n | \tfrac{\partial \varphi_n}{\partial R} \rangle \, dR$.

For only 1 parameter, if $R_1 = R_2$, the integral vanishes and $\gamma_n(t) = 0$.

If the Hamiltonian has more than 1 parameter, we have a path integral:

$$\gamma_n(t) = i\int_{\vec{R}_1}^{\vec{R}_2} \langle \varphi_n | \vec{\nabla}_R \varphi_n \rangle \cdot d\vec{R} \xrightarrow{\text{return to initial}} i\oint \langle \varphi_n | \vec{\nabla}_R \varphi_n \rangle \cdot d\vec{R} \quad \begin{pmatrix} \text{a closed} \\ \text{contour integral} \\ \text{in phase space} \end{pmatrix}$$

Return to spin-1/2

The eigenvectors ("spinors") of $H = \mu_B B(\hat{n}\cdot\vec{\sigma}) = \mu_B B \begin{bmatrix} \cos\theta & e^{-i\phi}\sin\theta \\ e^{i\phi}\sin\theta & -\cos\theta \end{bmatrix}$

are $|-\rangle = \begin{bmatrix} e^{-i\phi}\sin\theta/2 \\ -\cos\theta/2 \end{bmatrix}$ and $|+\rangle = \begin{bmatrix} \cos\theta/2 \\ e^{i\phi}\sin\theta/2 \end{bmatrix}$

The gradient in θ, ϕ parameter space is

$$\nabla_{\theta,\phi} = \frac{d}{d\theta}\hat{\theta} + \frac{d}{d\phi}\hat{\phi}$$

Thus we have for γ_-:

$$\nabla_{\theta,\phi}|-\rangle = \frac{1}{2}\begin{bmatrix} e^{-i\phi}\cos\theta/2 \\ \sin\theta/2 \end{bmatrix}\hat{\theta} + \begin{bmatrix} -ie^{-i\phi}\sin\theta/2 \\ 0 \end{bmatrix}\hat{\phi}$$

Note that the first term is $\frac{e^{-i\phi}}{2}|+\rangle$, orthogonal to $|-\rangle$.

So $\langle -|\nabla_{-,\phi}|-\rangle = -i\sin^2\frac{\theta}{2}\hat{\phi}$. Now, for our previous problem, $\theta = \alpha$ and one period path thru θ, ϕ phase space has "length" $2\pi\hat{\phi}$.

Therefore, $\gamma(T) = 2\pi \sin^2\frac{\alpha}{2}$, same as we found before!

Scattering from spherically symmetric potentials

In 3-D, the result of scattering from a localized potential $V(r)$ cannot be captured by transmission coef. as in 1-D, because the incident particle can be scattered not just backwards + forwards, but at all scattering angles θ. Instead, we want to calculate a quantity that indicates the effectiveness of the scatterer:

"differential scattering cross-section" $D(\theta) = \frac{d\sigma}{d\Omega}$.

With a classical trajectory:

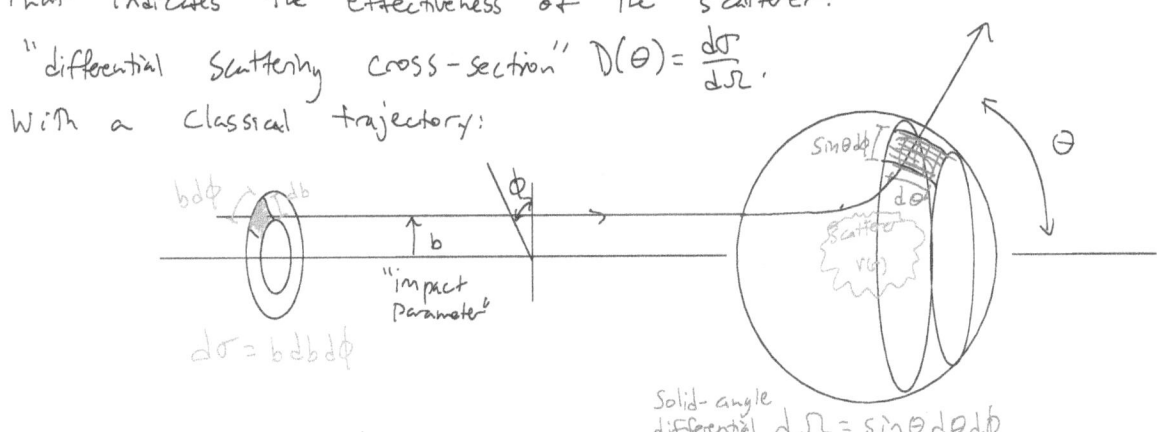

$d\sigma = b\,db\,d\phi$

Solid-angle differential $d\Omega = \sin\theta\,d\theta\,d\phi$

Clearly, $\frac{d\sigma}{d\Omega} = \frac{b\,db\,d\phi}{\sin\theta\,d\theta\,d\phi} = \frac{b}{\sin\theta}\left|\frac{db}{d\theta}\right|$ (added abs. val. to keep this positive)

Classical scattering from a hard sphere

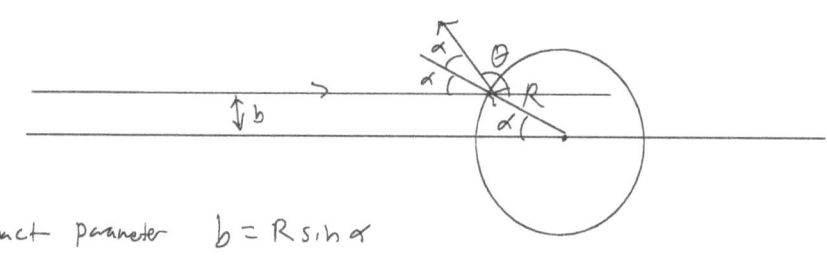

impact parameter $b = R\sin\alpha$

Scattering angle $\theta = \pi - 2\alpha$

$b = R\sin\left(\frac{\pi}{2} - \frac{\theta}{2}\right) = R\cos\frac{\theta}{2} \longrightarrow \theta = 2\arccos\frac{b}{R}$

So $D(\theta) = \frac{b}{\sin\theta}\left|\frac{db}{d\theta}\right| \xrightarrow{\frac{db}{d\theta} = -\frac{R}{2}\sin\frac{\theta}{2}} D(\theta) = \frac{R\cos\theta/2}{\sin\theta}\left(\frac{R\sin\theta/2}{2}\right) = \frac{R^2}{4}$

and **total** cross section $\sigma = \int D(\theta)\,d\Omega = 4\pi \cdot \frac{R^2}{4} = \pi R^2$

This is exactly what we expect: The sphere projects area of a circle w/ radius R in the plane transverse to particle motion!

Quantum mechanical wave scattering

$\Psi(r,\theta) = \Psi_{inc.} + \Psi_{scattered} = A\left\{e^{ikz} + f(\theta)\frac{e^{ikr}}{r}\right\}$, $k = \frac{\sqrt{2mE}}{\hbar}$

$f(\theta)$ contains all info about $\frac{d\sigma}{d\Omega}$:

$|\Psi_{inc.}|^2 \cdot dV = |A|^2 (vdt)\,d\sigma$, $|\Psi_{scatt}|^2 \cdot dV = |A|^2 \frac{|f|^2}{r^2} vdt\, r^2 d\Omega$

Because probability must be conserved, $d\sigma = |f|^2 d\Omega \longrightarrow \frac{d\sigma}{d\Omega} = |f|^2$

$\rightarrow$ find coef. of $\frac{e^{ikr}}{r}$ term in solution to radial Schrödinger Eqn. to determine $D(\theta)$!

Partial wave analysis

radial Schrödinger eqn: $-\frac{\hbar^2}{2m}\frac{d^2u}{dr^2} + V(r)u + \frac{\hbar^2 \ell(\ell+1)}{2mr^2}u = Eu$,

If scattering potential $V(r)$ is localized stronger than $\frac{1}{r^2}$ (Importantly, this does not apply to the bare Coulomb potential $\propto \frac{1}{r}$), radial eqn. is <u>approximately</u> $-\frac{\hbar^2}{2m}\frac{d^2u}{dr^2} + \frac{\hbar^2 \ell(\ell+1)}{2mr^2}u = Eu$ for <u>large</u> r.

This has out-going solutions called "spherical Hankel fns": $h_\ell^{(1)}(\kappa r)$ w/ asymptotic behavior $h_\ell^{(1)} \to \frac{1}{\kappa r}(-i)^{\ell+1} e^{i\kappa r}$.

So $\psi(r,\theta,\phi) = A\left\{ e^{ikz} + \sum_{\ell,m} C_{\ell m} h_\ell^{(1)}(\kappa r) Y_\ell^m(\theta,\phi) \right\}$

Since $V(r)$ and incoming wave do not break azimuthal symmetry, the solution cannot have ϕ dependence, and so only $m=0$ spherical harmonics can contribute:

$$Y_\ell^0(\theta,\phi) = \sqrt{\frac{2\ell+1}{4\pi}}\, P_\ell(\cos\theta) \quad \leftarrow \text{Legendre polynomial}$$

Partial wave amplitude

Setting $C_{\ell 0} = i^{\ell+1} \kappa \sqrt{4\pi(2\ell+1)}\, a_\ell$, where a_ℓ is "partial wave amplitudes", we can cancel some coefficients:

$$\psi(r,\theta) = A\left\{ e^{ikz} + \kappa \sum_{\ell=0}^{\infty} i^{\ell+1}(2\ell+1) a_\ell h_\ell^{(1)}(\kappa r) P_\ell(\cos\theta) \right\}$$

$$= A\left\{ e^{ikz} + \sum_{\ell=0}^{\infty} (2\ell+1) a_\ell P_\ell(\cos\theta) \frac{e^{ikr}}{r} \right\}$$

Evidently, $f(\theta) = \sum_{\ell=0}^{\infty} (2\ell+1) a_\ell P_\ell(\cos\theta)$, so that differential cross-section

$$D(\theta) = |f(\theta)|^2 = \sum_\ell \sum_{\ell'} (2\ell+1)(2\ell'+1) a_\ell^* a_{\ell'} P_\ell(\cos\theta) P_{\ell'}(\cos\theta)$$

Using orthogonality relation $\int P_\ell(\cos\theta) P_{\ell'}(\cos\theta)\, d\Omega = \frac{4\pi}{2\ell+1}\delta_{\ell\ell'}$,

total cross-section $\sigma = \int D(\theta)\, d\Omega = \sum_\ell 4\pi(2\ell+1)|a_\ell|^2$

Hard sphere scattering

Using plane wave expansion $e^{ikz} = \sum_l i^l (2l+1) j_l(kr) P_l(\cos\theta)$ ← Spherical Bessel fns.

$$\psi(r,\theta) = A \sum_l i^l (2l+1) \left(j_l(kr) + ik a_l h_l^{(1)}(kr) \right) P_l(\cos\theta)$$

So it was convenient to keep that factor of $2l+1$ in redefining C_{l0}!

Example: hard sphere scattering $V(r) \to \infty$ $r \leq R$. Then, boundary condition:

$$\psi(R,\theta) = \sum_l i^l (2l+1) \left[j_l(kR) + ik a_l h_l^{(1)}(kR) \right] P_l(\cos\theta) = 0$$

Since $P_l(\cos\theta)$ are orthonormal, $[\quad] = 0$ for all l, and $a_l = \frac{i j_l(kR)}{k h_l^{(1)}(kR)}$.

For low-energy scattering, k small and $kR \ll 1$. Then,

$$\frac{j_l(z)}{h_l^{(1)}(z)} = \frac{j_l(z)}{j_l(z) + i n_l(z)} \approx -i \frac{j_l(z)}{n_l(z)} \approx \frac{i}{2l+1} \left[\frac{2^l l!}{(2l)!} \right]^2 z^{2l+1}$$

$z = kR$ small, so $l=0$ term dominates and $\sigma \approx 4\pi |a_0|^2 = \frac{4\pi}{k^2}(kR)^2 = 4\pi R^2$.

Instead of seeing just projected area, the wave sees entire surface!

Green's function

$$\left(-\frac{\hbar^2}{2m}\nabla^2 + V\right)\psi = E\psi \implies (\nabla^2 + k^2)\psi = Q, \text{ where } k \equiv \frac{\sqrt{2mE}}{\hbar}, Q = \frac{2mV}{\hbar^2}\psi$$

If we can solve $(\nabla^2 + k^2) G(r) = \delta^3(r)$ for "Green's function" $G(r)$,

then solution to TISE is $\psi = \psi_0 + \int G(r-r_0) Q(r_0) d^3 r_0$:

$\underbrace{(\nabla^2+k^2)\psi_0}_{0 \text{ (homogeneous sol'n)}} + \int [(\nabla^2+k^2) G(r-r_0)] Q(r_0) d^3 r_0 = \int \delta^3(r-r_0) Q(r_0) d^3 r_0 = Q(r)$.

But what is $G(r)$? If $k=0$, then we have $\nabla^2 G = \delta^3(r)$.

This is Poisson eqn w/ point charge! In this case, we know $G = -\frac{1}{4\pi r}$.

For a screened point charge, Poisson eqn is $\nabla^2 \phi = C\phi$

with Green's fn $-\frac{e^{\sqrt{C} r}}{4\pi r}$. In TISE, $C \to -k^2$ so $G(r) = -\frac{e^{ikr}}{4\pi r}$.

and $\psi(\vec{r}) = \psi_0(r) - \int \frac{e^{ik|\vec{r}-\vec{r_0}|}}{4\pi |\vec{r}-\vec{r_0}|} \frac{2m}{\hbar^2} V(\vec{r_0}) \psi(\vec{r_0}) d^3 r_0 = \psi_0 - \frac{m}{2\pi\hbar^2} \int \frac{e^{ik|\vec{r}-\vec{r_0}|}}{|\vec{r}-\vec{r_0}|} V(\vec{r_0}) \psi(\vec{r_0}) d^3 r_0$

Born approximation

Far away, $r \gg r_0$ and
$$|\vec{r}-\vec{r_0}|^2 = r^2 + r_0^2 - 2\vec{r}\cdot\vec{r_0} = r^2\left(1+\left(\frac{r_0}{r}\right)^2 - 2\frac{\vec{r}\cdot\vec{r_0}}{r^2}\right) \approx r^2\left(1 - \frac{2\vec{r}\cdot\vec{r_0}}{r^2}\right)$$

so $|\vec{r}-\vec{r_0}| \approx r\left(1 - \frac{2\vec{r}\cdot\vec{r_0}}{r^2}\right)^{1/2} \approx r\left(1 - \frac{\vec{r}\cdot\vec{r_0}}{r^2}\right) = r - \frac{\vec{r}\cdot\vec{r_0}}{r} = r - \hat{r}\cdot\vec{r_0}$

and $\frac{e^{ik|\vec{r}-\vec{r_0}|}}{|\vec{r}-\vec{r_0}|} \approx \frac{e^{ik(r-\hat{r}\cdot\vec{r_0})}}{r-\hat{r}\cdot\vec{r_0}} = \frac{e^{ikr}e^{-i\vec{k}\cdot\vec{r_0}}}{r - \hat{r}\cdot\vec{r_0}} \approx \frac{e^{ikr}e^{-i\vec{k}\cdot\vec{r_0}}}{r}$ $(\vec{k} = k\hat{r})$

$$\Psi(r) = \Psi_0 + \left[-\frac{m}{2\pi\hbar^2}\int e^{-i\vec{k}\cdot\vec{r_0}} V(\vec{r_0})\Psi(\vec{r_0})d^3r_0\right]\frac{e^{ikr}}{r}$$

If $\Psi_0 = Ae^{ikz}$ and $V(r)$ is weak, then $\Psi(\vec{r_0}) \approx \Psi_0(\vec{r_0}) = Ae^{ikz_0} \xrightarrow{z_0 = \hat{z}\cdot\vec{r_0}} Ae^{i\vec{k}'\cdot\vec{r_0}}$, $\vec{k}' = k\hat{z}$

$$\Psi(r) \cong A\left(e^{ikz} + \underbrace{\left[-\frac{m}{2\pi\hbar^2}\int e^{i(\vec{k}'-\vec{k})\cdot\vec{r_0}} V(\vec{r_0})d^3r_0\right]}_{f(\theta) \rightarrow -\frac{m}{2\pi\hbar^2}\int V(\vec{r})d^3r \text{ (for small k)}}\frac{e^{ikr}}{r}\right)$$ "1st Born approx."

Spherically symmetric potential

Evaluate:
$$f(\theta) = -\frac{m}{2\pi\hbar^2}\int_0^\infty\int_0^\pi\int_0^{2\pi} e^{i|\vec{k}'-\vec{k}|r_0\cos\theta_0} V(r_0)r_0^2 \sin\theta_0 \, d\theta_0 \, d\phi_0 \, dr_0$$

$$= -\frac{2m}{\hbar^2 K}\int_0^\infty r V(r)\sin(Kr)dr, \text{ where } K = |\vec{k}'-\vec{k}|.$$

If we have elastic scattering, $|\vec{k}'| = |\vec{k}|$, and

$K = 2|\vec{k}|\sin\theta/2.$

law of sines:
$\frac{\sin\theta}{K} = \frac{\sin\alpha}{|\vec{k}|}$

$K = |\vec{k}|\frac{\sin\theta}{\sin\alpha},\quad \alpha = \frac{\pi-\theta}{2}$

$K = |\vec{k}|\frac{\sin\theta}{\cos\theta/2} = |\vec{k}|\frac{2\sin\theta/2\cos\theta/2}{\cos\theta/2}$
$= 2|\vec{k}|\sin\theta/2$

For $V(r) = -\frac{e^2}{4\pi\varepsilon_0 r}e^{-\mu r}$, $f(\theta) = \frac{2me^2}{4\pi\varepsilon_0\hbar^2 K}\int_0^\infty e^{-\mu r}\sin(Kr)dr = \frac{2me^2}{4\pi\varepsilon_0\hbar^2}\frac{1}{\mu^2 + K^2}$

In limit of $\mu \to 0$ (Coulomb) $D(\theta) = |f(\theta)|^2 \propto \frac{1}{K^4} \propto \frac{1}{\sin^4(\theta/2)}$

Identical to classical (Rutherford) result!

www.ingramcontent.com/pod-product-compliance
Lightning Source LLC
Chambersburg PA
CBHW080914170526
45158CB00008B/2104